STABILITY OF MOTION

Stability of Motion

APPLICATIONS OF LYAPUNOV'S SECOND METHOD TO
DIFFERENTIAL SYSTEMS AND EQUATIONS WITH DELAY

N. N. Krasovskii

S. M. KIROV URAL POLYTECHNIC INSTITUTE, SVERDLOVSK, R.F.S.F.R.

TRANSLATED BY

J. L. Brenner

STANFORD RESEARCH INSTITUTE

STANFORD UNIVERSITY PRESS · STANFORD, CALIFORNIA
1963

This book is a translation, with alterations and additions, of N. N. Krasovskiĭ, *Nekotorye zadači teorii ustoĭčivosti dviženiya* (Moscow, 1959)

Stanford University Press
Stanford, California

Library of Congress Catalog Card Number: 63-12040
Printed in the United States of America

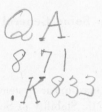

Contents

STABILITY OF MOTION

Introduction

1. Statement of the Problem

In this book we consider physical systems that can be described by ordinary differential equations in which the independent variable is time. In the last two chapters, the differential equations also involve a time delay. The theorems we shall state and prove concern stability (which we are going to define immediately) but are not necessarily connected with mechanical systems. The differential equations involved can be interpreted as coming from electrical or acoustical systems as well; in fact the same equation can be variously interpreted so that it describes several different systems. Thus our approach, which concerns the differential equations themselves, is a general one. The terms we use, "motion" and "stability of motion," are not to be narrowly interpreted.

Let t be the time, and suppose that some physical system is described by the variables y_i, where $y_i \equiv y_i(t)$, and that the conditions of motion require that these variables satisfy*

$$(1.1) \qquad \frac{dy_i}{dt} = Y_i(y_1, \cdots, y_n, t) \qquad (i = 1, \cdots, n) \, .$$

Suppose that $y_i = \eta_i(t)$ is some particular solution of (1.1). To study the properties of solutions of (1.1) in the neighborhood of the solution $\eta_i(t)$, we make the substitution

$$(1.2) \qquad x_i = y_i - \eta_i(t) \qquad (i = 1, \cdots, n) \, ,$$

and call $x_i = 0$, or $y_i = \eta_i(t)$, the unperturbed motion or unperturbed trajectory** (after Lyapunov [114, p. 209]). Since the variables y_i satisfy (1.1), the new variables x_i satisfy the differential equations

$$(1.3) \qquad \frac{dx_i}{dt} = X_i(x_1, \cdots, x_n, t) \qquad (i = 1, \cdots, n) \, ,$$

---

* In this book we are concerned only with real functions of real arguments.

** *Trajectory, motion,* and *solution* of the differential equations are used interchangeably in this book. *Orbit* means the locus described by the motion.

where

$$(1.4) \qquad X_i(x_1, \cdots, x_n, t) = Y_i(x_1 + \eta_1, \cdots, x_n + \eta_n, t) - Y_i(\eta_1, \cdots, \eta_n, t) .$$

Following Lyapunov, we call equations (1.3) the *equations of perturbed motion*. The variables y_i are suppressed in the remainder of the book, and variables x_i are treated directly. Thus the curve $\{\eta_i(t)\}$ is always denoted by $x_i = 0$, and is called the null solution (or the unperturbed solution).[*] As can be seen from (1.4), the functions X_i are assumed to have the property

$$(1.5) \qquad X_i(0, \cdots, 0, t) = 0 \qquad (i = 1, \cdots, n) .$$

As a rule, it will be assumed that these functions $X_i(x_1, \cdots, x_n, t)$ are continuous in some region G of $(x_1, \cdots, x_n)$-space for all positive values of the time $t \geqq 0$ (but see pp. 77–79). The region G will be an open set containing the point $x_1 = x_2 = \cdots = 0$; no supposition will be made concerning the boundedness of G. For many practical purposes it would be sufficient to consider only regions of the extremely simple form $\| x \| < H$ or $\| x \|_2 < H$, where H is a constant (or ∞) and where

$$(1.6) \qquad \| x \| = \sup (| x_1 |, \cdots, | x_n |) , \qquad \| x \|_2 = (x_1^2 + \cdots + x_n^2)^{1/2} .$$

With exceptions which are noted each time they occur, it is further assumed that in every closed bounded subregion $\bar{G}_\delta$, $\bar{G}_\delta \subset G$, the functions X_i satisfy a Lipschitz condition in their space arguments x_i; i.e.,

$$(1.7) \qquad | X_i(x_1'', \cdots, x_n'', t) - X_i(x_1', \cdots, x_n', t) | \leqq L_\delta \sup \{| x_j'' - x_j' |\} ,$$
$$L_\delta = \text{const.}, \qquad i = 1, \cdots, n .$$

This assures that solutions of (1.3) are unique in a neighborhood of $t = t_0$. The requirement that the solutions be unique is not an essential limitation, but is imposed merely to simplify the statement of the theorems.

In this book, the dependent quantities in the stability theorems are always denoted by x_i. When the theorems must be applied to a particular problem, such as the stability of a regulating mechanism, useful results can be obtained only if the dependent variables are properly chosen [114, p. 210].

As a matter of notation, the vector $x_1, \cdots, x_n$ is denoted by the single letter x. Similarly, the trajectory of (1.3) that takes on the values $x_i = x_{i0}$ at $t = t_0$ is denoted by $x_i(x_{10}, \cdots, x_{n0}, t_0, t)$, or simply by $x(x_0, t_0, t)$.

The fundamental definition (1.1, below) is that of Lyapunov. Some of the definitions of refined types of stability follow Četaev's annotations; see [36, pp. 11–12].

DEFINITION 1.1. *The null solution $x = 0$ of the system* (1.3) *is said to be stable (at $t = t_0$), provided that for arbitrary positive $\varepsilon > 0$ there is a $\delta = \delta(\varepsilon, t_0)$ such that, whenever $\| x_0 \| < \delta$, the inequality*

---

[*] A disturbed motion, or a perturbed trajectory, is one defined by other initial conditions. For the definition of persistent disturbances, see section 24, p. 100.

(1.8) $$\| x(x_0, t_0, t) \| < \varepsilon$$

is satisfied for all $t \geqq t_0$.

DEFINITION 1.2. *The null solution $x = 0$ of the system (1.3) is called asymptotically stable and the region G_δ of x-space is said to lie in the region of attraction of the point $x = 0$ (at $t = t_0$), provided that the conditions of definition 1.1 are satisfied, and provided further that*

(1.9) $$\lim_{t \to \infty} x(x_0, t_0, t) = 0 ,$$

(1.10) $$x(x_0, t_0, t) \in \Gamma, \qquad t \geqq t_0 ,$$

for all values of the initial point x_0 that lie in G_δ. Here Γ is some subregion of G which is given in advance, and with which the (mathematical model of the) physical problem is intrinsically concerned.

DEFINITION 1.2a. *The null solution $x = 0$ of the system (1.3) is called globally stable if the conditions of definition 1.2 are satisfied for every initial point x_0. In other words, the region of attraction of the point $x = 0$ must include the entire x-space.*

For applications, the problem is not the existence of the number δ and the region G_δ that correspond to a given region $\Gamma \subset G$ in definitions 1.1 and 1.2. Rather, it is a question of establishing suitable (and useful) bounds for these objects. We shall emphasize those methods which do give the possibility of finding these bounds. In this connection, it is important to classify *local stability*, as is done in the following definitions. *Stability in the large* is defined on p. 30 (definition 5.3).

DEFINITION 1.3. *The null solution $x = 0$ of the system (1.3) is stable for time intervals of length $T > 0$ for approximations $\Delta(\varepsilon) > 0$ if a modified condition (1.8) of definition 1.1 is satisfied, where, however, the words "for all $t \geqq t_0$" are replaced by the words "for $t_0 \leqq t \leqq T + t_0$," and the condition "$\delta > \Delta(\varepsilon)$" is appended.*

DEFINITION 1.4. *The null solution $x = 0$ of the system (1.3) is asymptotically stable at t_0 for a given interval of time $T > 0$ for approximations $\Delta(\varepsilon) > 0$, and $\gamma = \gamma(G_\delta, \theta) \geqq 0$, in Γ, provided that the conditions of definition 1.3 are satisfied, and provided that the further conditions*

(1.11) $$\| x(x_0, t_0, t) \| < \gamma , \quad for \quad t_0 + \theta \leqq t \leqq t_0 + T , \; \theta > 0 ,$$

(1.12) $$x(x_0, t_0, t) \in \Gamma , \quad for \quad t_0 \leqq t \leqq t_0 + T$$

are also satisfied for all initial points $x_0, x_0 \in G_\delta$. In the contrary case, we say that the null solution $x = 0$ is unstable in Γ for the given interval of time T, for the approximations in question.

There is no special significance in considering stability for a finite interval of time without including preassigned constants such as Δ, γ, and a region Γ in the definition. The fact is that the integrals in question are continuous; thus there is always a sufficiently small number $\delta > 0$ that satisfies the conditions of definition 1.3 for every fixed interval of time $t_0 \leqq t \leqq T$.

It was explicitly stated in definition 1.1, and should be emphasized, that the number δ can well depend not only on ε but also on the value of the initial moment of time t_0. Stability as defined by definition 1.1 is not a consequence of condition (1.9) but must be postulated in addition, as was done in definition 1.2. (See [83], [10].)

Definition 1.1 defines stability, and any motion that does not satisfy this definition is called unstable. The definition itself is motivated by applications, and it should be pointed out that an unstable motion is not necessarily an impractical one. This point is implicit in definitions 1.3 and 1.4. Further types of motion that are technically unstable but that may have practical importance will be described now.

First, it may be that practical interest resides in only certain coordinates $x_1, x_2, \cdots, x_k$, so that the criteria of the above definitions of stability, which require that all n coordinates remain small or approach 0, are too stringent. It may also be worthwhile to consider only certain functions $f_1, f_2, \cdots, f_k$ of the coordinates. (See the definition of *orbital stability* below.) The second sentence of this paragraph is an analytic paraphrase of the first under a suitable change of coordinates.

Second, it may be practically important to measure the set of initial values x_0, $\| x_0 \| < \delta$, for which the trajectories $x(x_0, t_0, t)$ remain in the vicinity of the null solution. The precise concept concerns the determination of the limit

$$m_\varepsilon = \lim_{\delta \to 0} \frac{m(S, \delta, \varepsilon)}{m(\delta)} ,$$

where $m(S, \delta, \varepsilon)$ is the measure of the set S of initial points x_0 for which not only $\| x_0 \| < \delta$, but also $\| x(x_0, t_0, t) \| < \varepsilon$; here ε is a positive number, and $m(\delta)$ is the measure of the sphere S_δ: $\| x_0 \| < \delta$. Further remarks and examples appear below.

Finally, we call attention to a definition. The null solution $x = 0$ of equations (1.3) is called *totally unstable* if there is a positive number ε such that for every x_0, $x_0 \neq 0$ in some sphere S_δ, the trajectory $x(x_0, t_0, t)$ eventually satisfies $\| x(x_0, t_0, t) \| = \varepsilon$. This definition clearly does not specify the worst kind of instability in any given case, and could be refined still further. Moreover, all the above definitions can be considered in regard to the dependence of the conditions in question on t_0, as well as the behavior during finite (and fixed) intervals of time. Extensive discussions appear in the books of Cesari [28] and Hahn [67], [68] and the survey paper of Antosiewicz [10]. Later in this book (see pp. 47–48), refinements of these definitions are given in which the stability in question (ordinary or asymptotic) is uniform with respect to a parameter. (See also the table on p. 61.)

The integral curves of a system of differential equations are parametrized by taking the initial values x_0 as parameters. This is by no means the only way to parametrize a family of curves, and the definition of stability can be generalized so that it applies to any family of curves that depends on a parameter or a set of parameters. This generalization is partly carried through in Chapter 5 of this book, and is explained by examples in the book [68].

The null solution of a system of differential equations is called *orbitally stable* if some functions (specified by the problem at hand) of the dependent variables change by only a small amount when the initial values of the dependent variables x_{i0} are restricted to a suitable neighborhood of 0. An example is the orbit of a planet moving in an inverse-square central force field. A slight change in position (or velocity, or both) may perturb the planet to an orbit with a different period. Hence two planets with initial positions and velocities that are nearly the same may eventually be very far apart, but the angular momenta, eccentricities, and other parameters that describe the orbits themselves remain close for all time.

A manifestation of orbital stability is seen in meteor showers. A possible explanation of meteor showers is based on a theory of origin of the family of meteors themselves. It is supposed that there was once a larger body which traveled in a definite orbit QQ, but which was broken up into smaller objects by a tidal force. At that moment, these objects had slightly different positions and velocities. The original motion in orbit QQ was unstable, so these tiny perturbations caused the family of small objects to be separated. The motion was orbitally stable, however; hence all the objects are spread around a fixed orbit. When the earth's orbit crosses the orbit QQ, which happens on a certain date, a meteor shower is observed. The precise definition is as follows.

DEFINITION 1.4a. *Let there be given $n-1$ independent, continuous functions $f_k(x_1, \cdots, x_n)$ of the arguments $x_i (k = 1, \cdots, n-1); f_k(0, \cdots, 0) = 0$. The null solution of (1.3) is called orbitally stable with respect to the orbit functions f_k, provided that a positive function $\delta = \delta(\varepsilon)$ of the positive real number ε exists so that $|f_k(x_1, \cdots, x_n)| < \varepsilon$ for all values of x_{i0} that satisfy $\| x_0 \| < \delta$.*

Remark. In certain cases, it may be desirable to modify this definition so that fewer than $n - 1$ functions are specified, or so that only those values of x come into question which satisfy some auxiliary condition. Orbital stability is important, for instance, if a picture of a fixed region on the earth's surface is to be taken from a satellite. (In rendezvous problems, the question seems to concern not orbital stability, but rather ordinary stability for a limited time T.)

Besides the above definitions of stability, more delicate ones have been considered. If the various restrictions $\| x(x_0, t_0, t) \| < \varepsilon$ hold not for all initial values satisfying $\| x_0 \| < \delta$ but only for almost all such values, the null solution may be unstable in the sense of Lyapunov, but stable from a practical point of view. The following situation may conceivably occur (although no concrete example can be adduced). The trajectories may be limited by the relation $\| x(x_0, t_0, t) \| < \varepsilon$ when $\| x_0 \| < \delta$ and all the coordinates x_{0i} are irrational, but not otherwise. In the case described the probability of a rational coordinate is zero, and even if one should occur, the physical problem would not be completely modeled by the differential system. Frequent small forces, such as collisions with micrometeorites, could disturb the motion, and convert an unstable trajectory into a stable one.

The following are a few simple examples of stable and unstable equilibrium. A solid homogeneous sphere resting on a horizontal flat table is stable. A small solid ball resting in a large rough spherical cup has a position that is asymptotically stable, the sense of this expression being physically obvious. If the size of the cup is constant in time, there is a uniformity in this stability, and the region of attraction is the (projection of the) diametral circle of the sphere. If the ball rests in an infinite cup (say $x^2 + y^2 = z$) that covers the entire (horizontal) xy-plane, the equilibrium position is *globally* asymptotically stable; this term is applied when the region of attraction is the entire physical space. A precise definition of global asymptotic stability appears on p. 3.

If the cup is a rough spherical one with radius $1/t$ which varies with time, the initial position is asymptotically stable, but not uniformly so in the initial value t_0. On the other hand, if the radius of the cup is $1/(1+t)$, the initial position is said to be asymptotically stable *in the large*. The terms *in the large* and *globally* are not synonymous.

A cone balanced on its vertex is totally unstable. A homogeneous wedge balanced on its apical line is unstable. This is true even if the wedge is braced by a weightless rod so it can fall to one side but not to the other. In this case, 50 percent of the initial positions in each circle $|| x_0 || \leq \delta$ lead to departing trajectories.

An example in which the null solution is unstable, but for which the limit m_ℓ is zero, can be constructed along the following lines. A cup is manufactured (Fig. 1) by connecting every point on the helical spiral

$$r = \theta \, ,$$
$$z = \theta \, , \qquad 0 \leq \theta < 2\pi$$

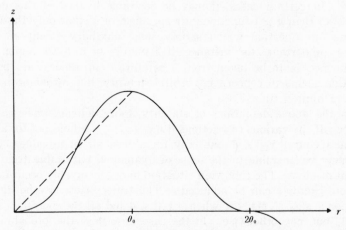

FIGURE 1. Section of a surface constructed to show that the measure of instability can be 0 for an unstable position.

to the origin by a straight-line segment $(r, \theta, z$ are cylindrical coordinates). These straight-line segments fill out a surface $z = f(x, y)$, which, however, is not defined for all values of x and y $(x = r \cos \theta, y = r \sin \theta)$, and which has a sharp corner at the origin. The trace of the surface in the plane $z = \theta_0$ is shown in the figure by a dotted line. If this trace is replaced by the solid curve, the new surface is indeed defined over the entire xy-plane, and is discontinuous only along the line $\theta = 0$ (except at $r = 0$). A final distortion of this surface in the neighborhood $2\pi - r^2 \leq \theta \leq 2\pi$ of this line of discontinuity gives a surface that is as smooth as desired. The null solution (a ball at rest at the apex) on this surface is unstable, but the measure m_ε of instability is 0.

Still another kind of stability that has been defined is *structural stability*. Most published definitions are general (imprecise), being refined and made precise when a particular problem is under consideration. If the right members of (1.3) are polynomials, the system is structurally stable if slight changes in the coefficients of the polynomials do not change the character of the differential system. A more general definition requires that the system be stable with respect to persistent disturbances. This question is considered in great detail in this volume. See the theorems on "structural properties" in Chapter 4.

A fundamental (and always available) method for studying the problem of stability is to construct a Lyapunov function. Lyapunov introduced this method in his book [114] and called it the *second method*. For a long time this method was considered to have only theoretical value, and was almost forgotten. The reversal of this situation is now so complete that Lyapunov's second method has even been applied to the design of specific control mechanisms.

By a Lyapunov function is meant a function $v(x_1, \cdots, x_n, t)$ defined in some region that contains the unperturbed solution $x = 0$ for all $t > 0$. Thus $v(x, t)$ is defined in an open region that contains the ray $x = 0, t > 0$ in x, t space. The Lyapunov theorems on stability concern conditions [114, pp. 259-66, theorems I–III] that the function v and its total derivative

$$(1.13) \qquad \frac{dv}{dt} = \sum_{i=1}^{n} X_i \frac{\partial v}{\partial x_i} + \frac{\partial v}{\partial t}$$

must satisfy. In principle, the method of Lyapunov makes it possible to obtain estimates for the values of the constants that occur in the definitions of stability given above. This book is concerned with stability problems that can be solved by Lyapunov's second method. If the function $v(x_1, \cdots, x_n, t)$ possesses certain properties (defined immediately below), it can be used to test whether the null solution of the differential system (1.3) is stable. The theorems which bear on this point require that the total derivative (1.13) enjoy certain properties as well. Formula (1.13) for the value dv/dt of this total derivative establishes the connection between the testing function $v(x_1, \cdots, x_n, t)$ and the system (1.3).

The solution vector with which we are concerned is henceforth denoted by $x_1, \cdots, x_n$.

The fundamentals of Lyapunov's second method are as follows. We consider a real function of the variables $x_1, \cdots, x_n, t$, denoted by $v(x_1, \cdots, x_n, t)$, or by $v(x, t)$, which is defined and continuous for all $t > 0$ in some region Γ of n-dimensional space. We also assume

$$(1.14) \qquad\qquad v(0, \cdots, 0, t) = 0 \quad \text{for} \quad t > 0 \,.$$

It is not required that the region Γ be a small neighborhood of the point $x = 0$, and when we consider stability for large initial perturbations, it cannot be.

DEFINITION 1.5. *If the inequality*

$$(1.15) \qquad\qquad v(x, t) \geqq 0 \,, \quad [or \quad v(x, t) \leqq 0]$$

holds for all x in Γ, and for all positive $t > 0$, the function $v(x, t)$ is said to be semidefinite in the region Γ.

DEFINITION 1.6. *Let $v(x_1, \cdots, x_n)$ be a function that does not depend explicitly on the time t. The function $v(x)$ is called definite in the region Γ if it is positive-definite or negative-definite in the region Γ; that is, if for all x in Γ, $x \neq 0$, the relation*

$$(1.16) \qquad\qquad v(x) > 0 \quad [or \quad v(x) < 0]$$

holds. The function $v(x, t)$ is called positive-definite [negative-definite] if

$$(1.17) \qquad\qquad v(x, t) \geqq w(x) \quad for \quad x \in \Gamma, \quad t > 0 \,,$$
$$[v(x, t) \leqq - w(x) \quad for \quad x \in \Gamma, \quad t > 0]$$

holds for some positive-definite function $w(x)$.

DEFINITION 1.7. *A function $v(x, t)$ admits an infinitely small upper bound in the region Γ (in extenso, admits an upper bound which is infinitely small at the point $x = 0$), provided that there is a continuous function $W(x)$ such that $W(0) = 0$ and the relations*

$$(1.18) \qquad\qquad | v(x, t) | \leqq W(x)$$

hold for $x \in \Gamma, t > 0$.

We state a theorem on stability and two theorems on instability. The proof of the theorem on stability is included. Converses to these theorems, and historical notes, are given in Chapter 1.

LYAPUNOV'S THEOREM ON STABILITY. [114, p. 259.] *Suppose there is a function v which is definite along every trajectory of (1.3), and is such that the total derivative dv/dt is semidefinite of opposite sign (or identically 0) along every trajectory of (1.3). Then the perturbed motion is stable. If a function v exists with these properties and admits an infinitely small upper bound, and* *if dv/dt is definite (with sign opposite to that of v), it can be shown further that every perturbed trajectory which is sufficiently close to the unperturbed motion x = 0 approaches the latter asymptotically.*

It should be emphasized that the theorem can be applied without solving the differential system. If the function v is positive-definite, and if its total derivative is negative-semidefinite, the hypotheses of the theorem are satisfied. To compute the total derivative of v, it is not necessary to have formulas for the trajectories, but only to know what the differential system is. [See (1.13).]

PROOF. It is supposed that $v(x, t)$ is continuous; without loss of generality, suppose $v(x, t) > w(x) > 0$ $(x \neq 0, 0 < t < \infty)$. Set $\lambda_\varepsilon = \inf W(x)$, the infimum being computed over the set $\| x \| = \varepsilon$. This infimum is positive by assumption. Since v is continuous, δ can be chosen so that $v(x_0, t) < \lambda_\varepsilon$ for $\| x_0 \| \leq \delta, t = t_0$. To see that the trajectory $x(x_0, t_0, t)$ $(\| x_0 \| < \delta)$ will remain in the region $\| x \| < \varepsilon$ for all time $t \geq t_0$, note that, since $dv/dt \leq 0$, relation $v(x, t) < \lambda_\varepsilon$ holds for $t \geq t_0$. This means that $\| x(x_0, t_0, t) \|$ never reaches the value ε, by the definition of λ_ε.

Suppose further that $v(x, t)$ admits an infinitely small upper bound: $v(x, t) < W(x)$, where $W(x)$ is continuous, $W(0) = 0$. Suppose also that $dv/dt < \alpha(x) < 0$, $x \neq 0$. It must be shown that the null solution is asymptotically stable. It has already been proven for every ε that the relation $\| x(x_0, t_0, t) \| < \varepsilon$ holds if $\| x_0 \| < \delta = \delta(\varepsilon)$. Suppose that $\| x(x_0, t_0, t) \| > \delta_1 > 0$ for all $t \geq t_0$. Then $dv/dt < -\alpha < 0$, and $v(x, t)$ would eventually become negative on every trajectory for which $\| x_0 \| < \delta$:

$$v(x, t) = v(x_0, t_0) + \int_{t_0}^{t} (dv/dt)\, dt < v(x_0, t_0) - \alpha(t - t_0) .$$

This contradiction shows that $x(x_0, t_0, t)$ eventually penetrates every neighborhood of the origin. Since $v(x(x_0, t_0, t), t)$ is a decreasing function of t, it follows that this function has limit 0, and $x(x_0, t_0, t) \to 0$.

COROLLARY. *If $v(x, t)$ is a function $v(x)$ that does not depend explicitly on the time and if X_i are similarly independent of t, the null solution is asymptotically stable.*

See Chapter 2, especially pp. 46–48, 62.

LYAPUNOV'S FIRST THEOREM ON INSTABILITY. [114, p. 262.] *Suppose a function v exists for which the total derivative dv/dt is definite along every trajectory of (1.3); suppose v admits an infinitely small upper bound; and suppose that for all values of t above a certain bound there are arbitrarily small values x_s of x for which v has the same sign as its derivative. Then the perturbed motion is unstable.*

LYAPUNOV'S SECOND THEOREM ON INSTABILITY. *(See p. 42.) Suppose a bounded function v exists for which the derivative dv/dt has the form*

$$(1.19) \qquad \frac{dv}{dt} = \lambda v + w$$

along the trajectories of (1.3), *where λ is a positive constant, and w is either identically zero or at least semidefinite. If w is not identically zero, suppose further that for all values of t above a certain bound there are arbitrarily small values x_s of x for which v and w have the same sign. Then the perturbed motion is unstable.*

These theorems find application in this book, especially in Chapters 4 through 7.

A particular differential system can be proved to be stable if a suitable function v can be exhibited or constructed. The construction of such functions v is the central problem for the application of Lyapunov's second method. It is remarkable that, in a satisfactorily large number of cases, a Lyapunov function v always exists when the differential system is stable or unstable (converse problem to Lyapunov's theorems). The first two chapters of this volume are concerned with the existence of Lyapunov functions.

A second question is the practical application of Lyapunov's method. No device could be expected to be completely successful, but in Chapters 4 through 7 general methods are given that are effective for a large class of systems.

It is by no means necessary to read the book in sequence to understand the material from the last chapters. Certain basic definitions are needed; they are contained in this introduction and are referred to by page number when they arise in specific applications. Aside from this, the chapters are entirely independent. Those who are interested in properties of Lyapunov functions and methods of constructing them will not need to read the first two chapters, which are devoted to proofs of the existence of such functions. A preliminary scanning of the table of contents is recommended to all readers.

Theorems on the Existence of a Function $v(x_1, \ldots, x_n, t)$
Satisfying the Conditions of Lyapunov's Theorems

2. Preliminary Remarks

In this chapter, we investigate the existence of a function $v(x_1, \cdots, x_n, t)$ satisfying the conditions of the classical theorems of Lyapunov's second method. We call a function of this type a Lyapunov function. As remarked in the Introduction, we can consider only the region in which the function v is defined. We make the following comments to explain the importance of the existence problem.

Lyapunov's theorems I, II, III [114, pp. 259–66], which were cited on pp. 8 and 9, give conditions sufficient for the asymptotic stability and instability of perturbed motion. (Here we consider only stability.) The theorems on the existence of Lyapunov functions do show that the method is universally applicable; if a motion is stable, this can always be established by using a Lyapunov function. Also, questions concerning the nature of the trajectory can be studied by exhibiting Lyapunov functions with appropriate properties.

Theorems on the existence of Lyapunov functions have still another application. One of the methods of solving the stability problem is to start from a given system and try to find a simpler system which approximates it, and for which the asymptotic stability or instability can be established, then to show that the corresponding property persists on returning to the original system. The importance of being able to construct a Lyapunov function for the auxiliary system, or of being able even to show that such a function exists, is clear. Theorems on the existence of Lyapunov functions can make it possible to prove that a property of a system is a consequence of the corresponding property of the linearized or variational equations. Indeed, recognition of this capability led Lyapunov to put forth his second method.

Not only can conditions for the existence of a Lyapunov function be established for ordinary differential equations but the method and its applications can be carried over to more general systems. One could hope that a method for proving the existence of a Lyapunov function might carry with it a constructive method for obtaining this function. This hope has not been realized. Even now there are no methods for demonstrating the existence of a Lyapunov function that are not (effectively) constructive methods.

The problem of the existence of Lyapunov functions first attracted attention about 1930. Fundamental work by the pupils of Četaev on the theory of stability appeared in the Kazan school. Additional quantitative work on the question of the existence of Lyapunov functions has been published by Malkin [116], [117], [124], [125]; Persidskiĭ [145]-[148]; Massera [129]-[131]; Barbašin [13], [14], [16], [17]; Kurcveĭl' [96], [97], [100]; Zubov [196]-[199]; Vrkoč [183]; Yoshizawa [192]; F. Brauer [25]; and others.

The existence of a Lyapunov function v with properties which guarantee that the trajectories be stable, asymptotically stable, or unstable is not only sufficient but also necessary. Thus if the trajectories are known to be stable, a Lyapunov function with appropriate properties must exist.

This makes it clear that the smoothness of the function v can be more decisive than the smoothness of the right members of the system (1.3) of perturbed differential equations.

In the remainder of the chapter we prove the existence of a function v in some neighborhood of a segment of a perturbed trajectory, and give conditions for the existence of a function v that possesses a derivative dv/dt of definite sign in the given region. Then we obtain as a consequence of the general result a transformation of Lyapunov's theorem on asymptotic stability and his first theorem on instability. Finally, we prove the converse of Lyapunov's second theorem on instability and Četaev's theorem on instability.

3. Some Useful Lemmas

LEMMA 3.1. *Let* G_0, H_0 *be chosen so that* $\bar{H}_0 \subset G_0, \bar{G}_0 \subset G, 0 \in G_0,$ *and* G_0 *is bounded. Let* T *have the property that for every* $\theta, 0 \leq \theta \leq T,$ *the entire solution arc* $x(x_0, t_0, t), 0 < t_0 \leq t \leq t_0 + \theta$ *of* (1.3) *lies in* G_0 *if* $x_0 \in H_0$. *Suppose the functions* X_i *satisfy a Lipschitz condition. Let* $\gamma > 0$ *be an arbitrary positive number. Constants* α, τ, d *and a function* $V(x_1, \cdots, x_n, t)$ *exist so that*

(a) *V is continuous and has continuous first-order partial derivatives in* $G,$ $-\infty < t < \infty.$

(b)

(3.1) $V(x, t) = 0$ *for* $|| x - x(x_0, t_0, t) ||_2 \geq \gamma, -\infty < t < \infty,$

(c)

(3.2) $\dfrac{dV}{dt} = \sum_{i=1}^{n} \dfrac{\partial V}{\partial x_i} X_i + \dfrac{\partial V}{\partial t} > d$

   for $|| x - x(x_0, t_0, t) ||_2 < \alpha, -\tau + t_0 \leq t \leq t_0 + \theta,$

(d)

(3.3) $\dfrac{dV}{dt} \geq 0$ *for* $-\infty < t < t_0 + \theta + \tau, || x || < \infty,$

   and for $|| x - x(x_0, t_0, t + \theta) ||_2 \geq \gamma, \quad t_0 + \theta + \tau \leq t_0 + \theta + 2\tau,$

(3.4) $V > d$ *for* $|| x - x(x_0, t_0, t) ||_2 < \alpha, \quad t_0 - \tau \leq t \leq t_0 + \theta,$

(3.5) $V = 0$ *for* $t < t_0 - 2\tau, \quad t > t_0 + \theta + 2\tau.$

The values of the numbers τ, α, d are independent of the choice of the point $x_0 \in H_0$; they depend only on the choice of the regions H_0 and G_0 and the numbers T, γ.

PROOF. The speed of motion of the point $x_i(t)$ along the phase trajectory $x_i(x_0, t_0, t)$ in the region G_0 is uniformly bounded; thus numbers $\tau > 0$ and $\eta > 0$ can be found so that the points x that satisfy the inequality

$$\| x - x(x_0, t_0, t) \|_2 < \eta$$

lie in the region G_0 for $-2\tau + t_0 \le t \le t_0 + \theta + 2\tau$, and in the region $\| x - x(x_0, t_0, t_0 + \theta) \|_2 < \gamma$ for $t_0 + \theta \le t \le t_0 + \theta + 2\tau$.

Let us construct polynomials $y(t) = [y_1(t), \cdots, y_n(t)]$, which satisfy the inequalities

(3.6) $$\| x(x_0, t_0, t) - y(t) \|_2 < \varepsilon , \qquad \left\| \frac{dx(x_0, t_0, t)}{dt} - \frac{dy}{dt} \right\| < \varepsilon ,$$

for $-2\tau + t_0 \le t \le t_0 + 2\tau + \theta$. By the Weierstrass theorem on approximation of continuous functions [136a], these polynomials always exist.* The value of the positive number ε in inequalities (3.6) is specified later.

The function $V(x, t)$ is defined as follows:

(3.7) $$V(x, t) = (t - t_0 + 2\tau)^p \exp\left[(t - t_0 + 2\tau)^{-1}(t - t_0 - \theta - 2\tau)^{-1}\right]$$
$$\cdot \exp\left(\{[\| x - y(t) \|_2^2 \exp 2q(t_0 - t)] - \beta^2\}^{-1}\right)$$

for

(3.8) $$\| x - y(t) \|_2 < \beta \exp q(t - t_0), \; -2\tau + t_0 < t < t_0 + \theta + 2\tau ,$$

(3.9) $$V(x, t) = 0 \text{ for other values of } x \text{ and } t .$$

Here

(3.10) $$q = 8n^2 L_0 , \quad \beta = \frac{\eta}{2} \exp\left[-q(T + 2\tau)\right] ,$$

where L_0 is the Lipschitz constant for the function X_i in the region G_0 ; and p is an integer satisfying the inequality

(3.11) $$p > \frac{(T + 4\tau)^2}{\tau^4} + 8q\left(1 + \frac{n}{\beta^2}\right)[\exp 2q(T + 2\tau)]^2 .$$

We assert that the function $V(x, t)$ satisfies all the conditions of the lemma if the positive number $\varepsilon > 0$ in inequality (3.6) is taken less than $\eta/2$ ($\varepsilon < \eta/2$) and sufficiently close to 0. The argument is given below.

(a) In the region $x \in G, -\infty < t < \infty$ the function V is continuous and has continuous derivatives of all orders. In the first place, this is obvious

---

* If the interval on which the approximation in question is desired is $[a, b]$, the approximating polynomial is the Bernstein polynomial

$$B_\mu(x) = \sum_{k=0}^{\mu} f\left(a + \frac{k}{\mu}(b - a)\right) C_k^\mu \frac{(x - a)^k(b - x)^{\mu-k}}{(b - a)^{2\mu}} ; \qquad \lim_{\mu \to \infty} B_\mu(x) \to f(x) .$$

See also [65b].

in the regions (3.8) and (3.9). It is not difficult to show that the function in the right member of (3.7) is a continuous differentiable function on the boundary of these regions. This amounts to the statement that the function $\varphi(r) = \exp[(r - a)^{-1}(r - b)^{-1}]$, $a < r < b$, has the desired properties for $r \to a + 0$ and $r \to b - 0$.

(b) If the numbers η, β, ε are properly chosen, it is evident that equalities (3.1) and (3.5) follow from (3.7) and (3.9).

(c) Inequality (3.2) is proved by calculating the derivative of the function $V(x, t)$ of (3.7) along a trajectory of the system (1.1).

In region (3.8) we find

$$(3.12) \qquad \frac{dV}{dt} = V \left(\frac{p}{t - t_0 + 2\tau} + \frac{\theta - 2(t - t_0)}{(t - t_0 + 2\tau)^2 (t - t_0 - \theta - 2\tau)^2} \right.$$

$$+ 2 \left\{ q \, || \, x - y(t) \, ||_2^2 - \sum_{i=1}^{n} [x_i - y_i(t)] \left(\frac{dx_i}{dt} - \frac{dy_i}{dt} \right) \right\} \exp 2q(t_0 - t)$$

$$\left. \cdot \{ || \, x - y(t) \, ||_2^2 \exp [2q(t_0 - t)] - \beta^2 \}^{-2} \right).$$

If $\varepsilon < \beta[\exp(-2q\tau)]/4$, inequality (3.11) shows that

$$\frac{dV}{dt} > \frac{pV}{2} \quad \text{for} \quad -\tau < t - t_0 < \theta + \tau, \quad || \, x - y(t) \, ||_2 < \frac{\beta}{2} \exp q(t - t_0).$$

Thus, if we suppose that ε is the smaller of the two quantities $\eta/2$, $\beta \exp(-2q)/4$, then relations (3.2) and (3.4) are satisfied by choosing $\alpha = (\beta/4) \exp[-2q(T + \tau)]$.

(d) We now turn to inequality (3.3). First, we consider the value of the quantity

$$(3.13) \qquad \psi = q \, || \, x - y(t) \, ||_2^2 - \sum_{i=1}^{n} [x_i - y_i(t)] \left(\frac{dx_i}{dt} - \frac{dy_i}{dt} \right)$$

for

$$(3.14) \qquad -2\tau \leq t - t_0 \leq \theta + \tau,$$

$$\frac{\beta}{2} \exp q(t - t_0) < || \, x - y(t) \, ||_2 < \beta \exp q(t - t_0).$$

We obtain the estimate

$$\left| \frac{dx_i}{dt} - \frac{dy_i}{dt} \right| \leq \left| \frac{dx_i}{dt} - \frac{dx_i(x_0, t_0, t)}{dt} \right| + \left| \frac{dx_i(x_0, t_0, t)}{dt} - \frac{dy_i}{dt} \right|$$

$$\leq | X_i(x, t) - X_i(x(x_0, t_0, t), t) | + \varepsilon \leq L_0 n \, || \, x - x(x_0, t_0, t) \, ||_2 + \varepsilon$$

$$\leq \varepsilon + n L_0 [|| \, x - y(t) \, ||_2 + || \, y(t) - x(x_0, t_0, t) \, ||_2]$$

$$\leq \varepsilon + n L_0 [\varepsilon + \beta \exp q(t - t_0)].$$

Thus, in region (3.14), the inequality

$$(3.15) \qquad \psi \geq \frac{q\beta^2}{4} \exp 2q(t - t_0) - [n\beta \exp q(t - t_0)] [\varepsilon + L_0 \varepsilon + L_0 \beta \exp q(t - t_0)]$$

is satisfied. Since q satisfies the relation (3.10), we see that if $\varepsilon > 0$ is a sufficiently small number, relation (3.3) will also be satisfied in region (3.14), in view of the inequality (3.15).

Thus we have shown that the function V of (3.7), (3.9) satisfies all conditions (3.1)–(3.5) of the lemma. The numbers α, d, and τ do not depend on the arc $x(x_0, t_0, t)$ in the region H_0. This is clear because only properties of the functions X_i in the region G_0, and the value of the number T, were used to specify the values of the numbers η, τ, q, p, β, and ε. The lemma is proved.

The following is a corollary to lemma 3.1.

LEMMA 3.2. *Let G_0, H_0 be chosen so that $\bar{H}_0 \subset G_0, \bar{G}_0 \subset G, 0 \in G_0$, and G_0 is bounded. Let T have the property that for every $\theta, 0 \leq \theta \leq T$, the entire arc $x(x_0, t_0, t)$, $t_0 - \theta \leq t \leq t_0$ of a trajectory of (1.3) lies in G_0 if $x_0 \in H_0$. Suppose the functions X_i satisfy a Lipschitz condition. Let $\gamma > 0$ be an arbitrary positive number. Constants α, τ, d and a function $V(x, t)$ exist so that:*

(a) *V is continuous and has continuous first-order partial derivatives in G, $-\infty < t < \infty$,*

(b)

(3.16) $V(x, t) = 0 \quad for \quad \| x - x(x_0, t_0, t) \|_2 \geq \gamma, \quad -\infty < t < \infty$,

(c)

(3.17) $\dfrac{dV}{dt} > d \quad for \quad \| x - x(x_0, t_0, t) \|_2 < \alpha, \quad -\theta + t_0 \leq t \leq t_0 + \tau$,

(3.18) $\dfrac{dV}{dt} \geq 0 \quad for \quad -\tau - \theta + t_0 < t < \infty, \quad \| x \| < \infty$,

(d)   *and for* $\| x - x(x_0, t_0, \theta) \|_2 \geq \gamma, \quad t_0 - \theta - 2\tau \leq t \leq t_0 - \theta - \tau$,

(3.19) $V < -d \quad for \quad \| x - x(x_0, t_0, t) \|_2 < \alpha, \quad -\theta + t_0 \leq t \leq t_0 + \tau$,

(e)

(3.20) $V = 0 \quad for \quad t \leq t_0 - \theta - 2\tau, \quad t \geq t_0 + 2\tau$.

The values of the numbers τ, α, and d are independent of the choice of the point $x_0 \in H_0$; they depend only on the choice of the regions H_0, G_0, and the values of T and γ: $-N_0 < \partial V / \partial x_i, \partial V / \partial t < N_0$.

Lemma 3.2 follows from lemma 3.1 on making the substitutions $t \to -t$, $V \to -V$. It is clear that the supposition $t_0 \geq \theta$ was made in lemma 3.2 so that the trajectory $x(x_0, t_0, t)$ would be defined for $t \in [t_0 - \theta, t_0]$.

Remarks. (a) It follows from the proof of lemma 3.1 that the first partial derivatives $\partial V / \partial x_i, \partial V / \partial t$ are uniformly bounded by some constant N_0, and the value of N_0 depends only on the regions H_0, G_0, and the numbers T and γ.

(b) Let it be supposed that the functions X_i are uniformly continuous in the time t in every region G_0 in x-space, $0 \leq t < \infty$. Then the functions $x(x_0, t_0, t)$ and $dx_i/dt, i = 1, \cdots, n$, are bounded and equicontinuous functions of the time t for $t_0 - 2\theta \leq t \leq t_0 + \theta + 2\tau$ [or for $t_0 = \theta - 2\tau \leq t \leq t_0 + 2\tau$]. In making this statement, we consider only segments of the trajectory $\{x_i(x_0, t_0, t)\}$ that lie entirely in the region G_0 for the specified intervals of time. Under these hypotheses, whenever a positive number $\varepsilon > 0$ is given arbitrarily, we can find a *finite* system of polynomials $\{y_i(t)\}$ with the following property. For trajectory segments $x(x_0, t_0, t)$ restricted as above, there will be polynomials from this finite system that approximate the chosen segment of the trajectory in the sense of relation (3.6). Indeed, a family of uniformly bounded and equicontinuous functions admits a finite ε-set of polynomials. [This is established as follows. First the limit function is approximated to within $\varepsilon/4$ by a suitable polynomial. Because of the uniform convergence, all but a finite number of the functions in the sequence are approximated to within $\varepsilon/2$ by this same polynomial. Each of the finite number of functions remaining to be approximated is approximable to within ε by its private polynomial because of the equicontinuity.] Thus formula (3.7) shows that there is a finite set of functions $\{V\}$ such that on every trajectory segment $x(x_0, t_0, t)$ of the above type we can find a function V of this set that satisfies the conditions of lemma 3.1 (or of lemma 3.2). It follows that for every k we can find a constant N_k so that there is a function V which satisfies not only the conditions of lemma 3.1 (or lemma 3.2) but also the inequalities

$$\left| \frac{\partial^k V}{\partial x_1^{k_1} \cdots \partial x_n^{k_n} \partial t^{k_{n+1}}} \right| < N_k \qquad (k = 1, 2, \cdots).$$

4. Conditions for the Existence of a Lyapunov Function Having a Derivative dv/dt of Definite Sign

By a Lyapunov function we mean a testing function $v(x_1, \cdots, x_n, t)$ that is definite, and that possesses certain other properties that are hypotheses of the various theorems on stability. If the only converse theorem were a theorem guaranteeing the existence of a Lyapunov function with semidefinite total derivative dv/dt, the second method would be a weak one of only limited applicability. Fortunately, valid converses of several of the stability and instability theorems can be proved (existence of Lyapunov functions with derivatives that are definite in sign).

We recall that Lyapunov calls the function dv/dt positive-definite in a region Γ if there is a continuous function $w(x_1, \cdots, x_n)$, so that the conditions

$$(4.1) \qquad \frac{dv}{dt} = \sum_{i=1}^{n} \frac{\partial v}{\partial x_i} X_i + \frac{\partial v}{\partial t} \geqq w(x_1, \cdots, x_n),$$

$$(4.2) \qquad w(x_1, \cdots, x_n) > 0 \quad \text{for} \quad x \neq 0$$

are satisfied in the region Γ, and negative-definite if the inequality signs are reversed.

The hypothesis that the derivative dv/dt should have a definite sign is involved in two of Lyapunov's theorems, the theorem on asymptotic stability [114, p. 261] and the first theorem on instability [114, p. 262]. It will be recalled also (see pp. 8–9) that a further requirement of these theorems is that the function admit an infinitely small upper bound. Lyapunov [114,

p. 258] says that the function $v(x, t)$ admits an infinitely small upper bound in the region Γ if a continuous function $W(x)$ exists for which the conditions

(4.3) $$W(0) = 0, \quad |v(x, t)| \leq W(x)$$

are satisfied for $x \in \Gamma$.

In this section, we shall prove a general theorem on the existence of a Lyapunov function that has a derivative of definite sign in a given region. In the succeeding sections we shall use this general theorem to discuss the existence of a function v satisfying the conditions of these Lyapunov theorems. We note that the theorems of Lyapunov demand only that the function v be continuous; they do not impose more rigid requirements on the smoothness of this function. However, practical application requires that the function v have a derivative that can be represented in the form

(4.4) $$\frac{dv}{dt} = \sum_{i=1}^{n} \frac{\partial v}{\partial x_i} X_i + \frac{\partial v}{\partial t} ,$$

where equations (1.3) have been used. Thus, it is desirable to demonstrate the existence of a function v that has continuous first-order partial derivatives. There are problems in which it is important to know also that the second-order partial derivatives of v exist and are continuous. (See, for example, pp. 104-6.) Thus, in theorems on the existence of the function v, it is usually necessary to have a guarantee that v is sufficiently smooth.

The definition of asymptotic stability (p. 3, definition 1.2) included an estimate of the region of attraction G_δ. An alternative formulation of Lyapunov's Theorem on Asymptotic Stability is the following, which includes an estimate of the region G_δ in the formulation of the theorem.[*]

THEOREM 4.1. *Let Γ be a region that is contained in the region G and contains the point $x = 0$. If the differential equations of perturbed motion admit a testing function v such that*

(i) *v is positive-definite in Γ,*
(ii) *v admits an infinitely small upper bound in the region Γ,*
(iii) *dv/dt is a negative-definite function in the region Γ,*
(iv) *there are bounded regions G_δ, H_0, $G_\delta \subset H_0$, $\bar{H}_0 \subset \Gamma$, such that*

(4.5) $$\sup [v(x_1, \cdots, x_n, t) \text{ on the boundary of } G_\delta \text{ for } 0 \leq t < \infty]$$
$$< \inf [v(x_1, \cdots, x_n, t) \text{ on the boundary of } H_0 \text{ for } 0 < t < \infty],$$

then the unperturbed motion $x = 0$ is asymptotically stable and the region G_δ lies in the region of attraction of the point $x = 0$.

In the case of instability it is of practical importance to have an estimate for those regions which are forsaken by an unstable perturbed trajectory as time increases. The Lyapunov theorem on instability can be restated in the following form: if there is a region Γ that contains the point $x = 0$,

---

[*] As Četaev mentions [36, p. 16], this estimate is essentially given by Lyapunov in his proof of the theorem.

and the function v enjoys the properties enumerated in the theorem above, the perturbed trajectory $x(x_0, t_0, t)$ will leave an arbitrary bounded closed region $\bar{H}_0 \subset \Gamma$, provided sign $v(x_0, t_0, t) = $ sign dv/dt in Γ.

The proof of these theorems is given in detail by Lyapunov [114]. (See p. 64 of this volume.) The validity of the alternative formulation we have given is clearly established by Lyapunov's own proofs.*

The existence of the function v that satisfies the conditions of the theorems above is a basic question. If the function has a derivative dv/dt of definite sign, we can show that the character of the trajectories of perturbed motion does not change if the equations themselves change by a sufficiently small amount. It is thus possible to investigate problems of stability by simplifying or approximating the equations in question.

The existence of a Lyapunov function v that satisfies conditions of the theorems on asymptotic stability has been studied by a number of authors. The first results of a general nature were obtained by Massera [129]–[131], who proved that if the functions X_i in the right member of equations (1.3) of perturbed motion are periodic in time and continuously differentiable, then there exists a continuously differentiable Lyapunov function $v(x, t)$ in a neighborhood of the asymptotically stable unperturbed motion $x = 0$. Barbašin proved in [13], which appeared at the same time as Massera's work, that if the functions $X_i(x)$ have continuous partial derivatives up to the mth order inclusive $(m \geq 1)$, there exists a Lyapunov function $v(x)$ which possesses continuous partial derivatives up to the mth order inclusive.

For the case in which the right members $X(x, t)$ of the equations of perturbed motion are continuously differentiable, Malkin [125] gave necessary and sufficient conditions for the existence of a continuously differentiable Lyapunov function in some neighborhood of an asymptotically stable unperturbed trajectory. References [16] and [17] contain conditions that there exist a Lyapunov function $v(x, t)$ throughout phase space $-\infty < x_i < \infty$ $(i = 1, \cdots, n)$ for global stability (i.e., asymptotic stability of the solution $x = 0$, the region of attraction G_δ being the entire space). Extremely interesting results were obtained by Kurcveil' [97] and Massera [131], who showed that if the functions X_i in the equations of perturbed motion are only continuous, there does exist a Lyapunov function v, which is as smooth as desired, if the solution is asymptotically stable. There are also theorems on the existence of a Lyapunov function for dynamical systems in abstract spaces [91], [93], [196], [197], [199].

The existence of a function $v(x, t)$ satisfying the conditions of Lyapunov's first theorem on instability was considered by the author in [84] and [89].

We shall now adapt the known methods of proving the existence of a Lyapunov function v in the case of asymptotic stability. Theorems on the existence of a Lyapunov function will be proved by a unified method based on lemmas 3.1 and 3.2. We can use the method for both the case of stability and the case of instability.

We adopt the following notation. As always, let the functions X_i of the

---

* See preceding footnote.

right. members of equations (1.3) be defined in a region G and satisfy conditions (1.5) and (1.7) there.   Let Γ be a bounded region containing the point $x = 0$.   The symbol $\rho[p, q]$ denotes the distance between the points x_p, x_q:

$$\rho[p, q] = \| x_p - x_q \|_2 .$$

The symbol $\rho[P, Q]$ denotes the distance between sets P and Q.   The symbol O denotes the point $x = 0$.   We set

$$h = \min \rho[\bar{\Gamma} \backslash \Gamma, O] .$$

We choose (for later convenience) a set of monotonically decreasing numbers $h_0 = h, h_1, h_2, \cdots$, so that $\lim h_k = 0$ $(k \to \infty)$.   By $R(k)$ we mean the point set consisting of points for which $\| x \|_2 < h_k$.

We are interested in conditions under which there is a Lyapunov function v that has a derivative dv/dt of definite sign in the given region and that admits an infinitely small upper bound.   We shall see that the existence of such a function is closely connected with a certain property of the equation of perturbed motion; this we shall call *property* A.

DEFINITION 4.1.   *Let $h_k, R(k)$ be defined as above.   Suppose that for every closed bounded region $\bar{H}_0, \bar{H}_0 \subset \Gamma$, and for every positive integer $k, k > 0$, there is a number T_k such that whenever $t_0 \geq T_k, x_0 \in \Gamma \backslash R(k)$, the segment $t_0 - T_k \leq t \leq t_0 + T_k$ of the trajectory $x(x_0, t_0, t)$ does not lie entirely in the region H_0.   Then we say that region Γ has property* A.

We show first that property A is a necessary condition for the existence of a function v that satisfies Lyapunov's theorem on asymptotic stability and his first theorem on instability.

THEOREM 4.2.   *Suppose that for every bounded region $H_0, \bar{H}_0 \subset \Gamma$, there is a function $v(x_1, \cdots, x_n, t)$ that admits an infinitely small upper bound in H_0 and has a derivative dv/dt that is positive-definite or negative-definite in this region (along a trajectory of the equations (1.3) of perturbed motion).   Then the region Γ has property* A.

PROOF.   Let $R(m)$ be a neighborhood of the point $x = 0$ that (with its closure) lies entirely interior to H_0.   Let $x_0 \in H_0 \backslash R(m)$, and suppose $v(x_0, t_0) \geq 0$.   (If $v < 0$, see below.)   After time $\Delta t = \vartheta_m$, the trajectory $x(x_0, t_0, t)$ lies outside the sphere

(4.6) $$S(m): \| x \|_2 < R(m) \exp(-n L_0 \vartheta_m)$$

because of inequality (4.11), where ϑ_m is a fixed positive number.   [Inequality (4.11) is proved below.]   Here L_0 is the Lipschitz constant for the functions X_i in the region H_0.

Let $\bar{H}_1 \subset H_0$.   The value of the derivative dv/dt along a trajectory of (1.3) has a positive minimum $\delta_m > 0$, in $\bar{H}_1 \backslash S(m)$.   Thus, whenever a segment of the trajectory $x(x_0, t_0, t)$ lies inside the region $\bar{H}_1$ for all $t, t_0 \leq t \leq t_0 + \vartheta_m$, we see that

(4.7) $$v\big(x(x_0, t_0, t_0 + \vartheta_m), t_0 + \vartheta_m\big) > \vartheta_m \delta_m .$$

The function $v(x, t)$ admits an infinitely small upper bound. Hence we can find a neighborhood

(4.8) $$\| x \|_2 \leq h_{N(m)}$$

in which the inequality

(4.9) $$| v(x, t) | < \vartheta_m \delta_m$$

is satisfied. Moreover, the value of dv/dt is positive, $dv/dt > 0$, for the interval of time in which the trajectory $x(x_0, t_0, t)$ remains in the region H_0. Thus we conclude from (4.7) and (4.9) that a trajectory $x(x_0, t_0, t)$ which lies in the region H_0 cannot enter the sphere (4.8) as t increases. Moreover, outside the sphere (4.8) the inequality $dv/dt > \eta_m$ is satisfied along a trajectory in $\bar{H}_1$, where η_m is a positive constant. Thus the inequality

(4.10) $$v(x(x_0, t_0, t), t) > \vartheta_m \delta_m + (t - \vartheta_m)\eta_m$$

is satisfied for $t > \vartheta_m$. Set $m_1 = \max | v |, x \in \bar{H}_1, T_m = M_1/\eta_m + \vartheta_m$. It follows from (4.10) that the segment $t_0 \leq t \leq t_0 + T_m$ of the trajectory $x(x_0, t_0, t)$ cannot lie entirely in the region $\bar{H}_1$.

If $v(x_0, t_0) < 0$, a corresponding argument can be applied to the function $v(t) = v(x(x_0, t_0, t), t)$, as t decreases. If we change t to $-t$ in the arguments above, we find that the segment of $x(x_0, t_0, t)$ for which $t_0 - T_m \leq t \leq t_0$ cannot lie entirely in $\bar{H}_1$. This completes the proof of the theorem, since $\bar{H}_1$ is an arbitrary subregion of $H_0 \subset \varGamma$. If the functions $X(x, t)$ in the right member of equations (1.3) happen to be periodic functions of time, all with the same period ϑ (or in case these functions X are independent of the time), there is a clear geometric interpretation of property A.

LEMMA 4.1. *Let the functions X_i be periodic functions of the time t, all with the same period, or independent of time.*

A necessary and sufficient condition that the region $\varGamma$ enjoy property A is that whenever H_0 is bounded and $\bar{H}_0 \subset \varGamma$, no complete trajectory $x(x_0, t_0, t)$, $-\infty < t < \infty$ (except the point $x = 0$) lie entirely in H_0.

PROOF. The necessity of the condition in the lemma is obvious. To demonstrate sufficiency, let $R(m)$ be a sphere lying (with its closure) in the arbitrary bounded region $H_0, \bar{H}_0 \subset \varGamma$. Suppose *per contra* that there is a sequence of points $x_0^{(l)} \in \bar{H}_0 \backslash R(m)$ and a sequence of moments of time $t_0^{(l)}$ such that segments of the trajectories $x(x_0^{(l)}, t_0^{(l)}, t)$ lie entirely in the region H_0 for all $t_0^{(l)} - T_l \leq t \leq t_0^{(l)} + T_l$, and that $T_l \to \infty$ as $l \to \infty$. Since the set $H_0 \backslash R(m)$ is a compact set, it is possible to find a subsequence of the $x_0^{(l)}$ that has a limit point $x_0 \in \bar{H}_0$; suppose $x_0^{(l)}$ is already this subsequence, and that (by the same argument) the moments of time $t_0^{(l)}$ also have a limit $t_0 \in [0, \vartheta)$. Indeed, since the functions X_i are periodic, the initial moments of time t_0 can all be assumed to lie in the interval $[0, \vartheta)$. It is clear that $x_0 \neq 0$. Let $\bar{H}_1$ be a bounded region with $\bar{H}_1 \subset \varGamma$ and $\bar{H}_0 \subset H_1$. Since the trajectory $x(x_0, t_0, t)$ cannot lie entirely in the region H_1, there must be a moment of time $t = t_1$ for which $x(x_0, t_0, t) \in \bar{H}_1 \backslash H_1$. On the other hand, the integral curves are continuous; hence for l sufficiently large, the points $x(x_0^{(l)}, t_0^{(l)}, t_1)$ enter an arbitrarily small neighborhood of the point $x(x_0, t_0, t_1)$. Thus the latter point must lie outside the region $\bar{H}_0$. This contradicts the choice of the points $x_0^{(l)}$, and this contradiction establishes the lemma.

Thus we have proved that property A is a necessary condition for the existence of the Lyapunov function v. The study of the converse to Lyapunov's theorem can thus be reduced to the question whether property A is satisfied. We shall now show that property A is not only a necessary, but also a sufficient condition for the existence of a function v that admits an infinitely small upper bound and has a derivative of definite sign.

THEOREM 4.3. *If the region* Γ, $\Gamma \subset G$, *has property* A, *then every bounded region* H_0 *such that* $\bar{H}_0 \subset \Gamma$ *enjoys the following properties.*

(i) *There is a function* $v(x_1, \cdots, x_n, t)$ *defined in the region* H_0, *such that* dv/dt (4.4) *has a definite sign along a trajectory of* (1.3); $v(x_1, \cdots, x_n, t)$ *has partial derivatives* $\partial v/\partial x_i, \partial v/\partial t$, *which, together with* v, *are continuous and uniformly bounded in* H_0 *for all* $t, 0 < t < \infty$.

(ii) *Suppose the functions* X_i *to be uniformly continuous with respect to the time in the region* Γ, $0 < t < \infty$. *Then the function* $v(x_1, \cdots, x_n, t)$ *has continuous partial derivatives of all orders with respect to all arguments. Each derivative of arbitrary order*

$$\frac{\partial^m v}{\partial x_1^{m_1} \cdots \partial x_n^{m_n} \partial t^{m_{n+1}}}$$

is uniformly bounded in the region $H_0, 0 < t < \infty$ (*the bound being peculiar to the particular derivative*).

(iii) *If the functions* X_i *are periodic functions of the time* t, *all with the same period* ϑ (*or if the functions* X_i *are independent of the time* t), *then if the region* H_0 *has property* A *there is a function* $v(x_1, \cdots, x_n, t)$ *that is periodic in the time* t *with period* ϑ (*or independent of the time* t).

PROOF. We use the following notation. Let γ be a number such that $3\gamma < \rho[\bar{\Gamma}\backslash\Gamma, H_0]$, and let $H_0(\gamma)$, $H_0(2\gamma)$ be the set of points x that satisfy the relations $\rho[x, H_0] < \gamma, \rho[x, H_0] < 2\gamma$ respectively.

Let h_k be the first number in the sequence h_m defined above (p. 19) that satisfies the relation $h_k < \rho[\bar{H}_0\backslash H_0, 0]$.

Let $F_m = H_0\backslash R(m)\,(m \geq k)$. Let T_m be as defined in definition 4.1. First we shall show that for every integer $m \geq k$, there is a number $N(m)$ such that whenever $x_0 \in F_m$, a segment of the trajectory $x(x_0, t_0, t)$ that lies in the region Γ for $t_0 \leq t \leq t^*$ (or for $t^* \leq t \leq t_0$), $|t^* - t_0| \leq T_m$, has no point interior to the region $R(N(m))$. Indeed, the inequality $|x_i(dx_i/dt)| = |x_iX_i| < L_0 \|x\|_{\frac{1}{2}}^2$ is satisfied in the region $\bar{H}_0$ along the trajectory $x(x_0, t_0, t)$, where L_0 is the Lipschitz constant for the functions X_i in the region H_0. Summing on i, we obtain the differential inequality

$$d(\|x\|_{\frac{1}{2}}^2 \leq 2nL_0 \|x\|_{\frac{1}{2}}^2 |dt| = 2Q \|x\|_{\frac{1}{2}}^2 |dt|, \qquad Q = \text{const.}$$

By integrating the latter inequality, we obtain

(4.11) $|| x_0 ||_2 \exp(-Q | t - t_0 |) \leq || x(x_0, t_0, t) ||_2 \leq || x_0 ||_2 \exp(Q | t - t_0 |)$.

Now if $x_0 \in F_m$, it follows that $|| x ||_2 \geq h_m$, and consequently $|| x(x_0, t_0, t) || \geq$ $h_m \exp(-QT_m)$. This establishes our assertion if we choose $N(m)$ as the smallest integer that satisfies the inequality $h_{N(m)} < h_m \exp(-QT_m)$. Let γ_m be a number that satisfies the inequalities $\gamma_m \leq \frac{1}{2}[h_{N(m)} - h_{N(m)+1}]$, $\gamma_m \leq \gamma$.

Now we turn to construction of the function $v(x, t)$. Consider a point $x_0 \in F(m)$, $m \geq k$. Property A shows that we can choose a number θ such that whenever $| \theta | \leq T_m$ (or $t_0 < T_m$), we have $x(x_0, t_0, t + \theta) \in [\bar{H}_0(2\gamma) \backslash H_0(2\gamma)]$. For definiteness, let $\theta > 0$. Then, by lemma 3.1, there is a function $V = v(x_0, t_0, x, t)$, defined and continuous in the region G for all values of the time t. The function $v(x_0, t_0, x, t)$ has continuous and uniformly bounded partial derivatives $\partial v/\partial x_i$, $\partial v/\partial t$. (See remark (a), p. 15.) If we take the number $\gamma = \gamma_m$ in lemma 3.1, we see that the following conditions are satisfied for $t_0 - \tau_m < t < t_0 + \tau_m$:

(4.12) $v(x_0, t_0, x, t) = 0$ in the region $|| x ||_2 \leq h_{N(m)+1}$,

(4.13) $\dfrac{dv(x_0, t_0, x(t), t)}{dt} \geq 0$ in the region H_0 ,

(4.14) $\dfrac{dv(x_0, t_0, x(t), t)}{dt} > d_m$ in the region $|| x - x(x_0, t_0, t) ||_2 < d_m$,

where the numbers $\alpha_m > 0, \tau_m > 0, d_m > 0$ depend only on the number m, and are independent of the choice of the point $x_0 \in F_m$. The symbol $dv(x_0, t_0, x(t), t)/dt$ denotes the total derivative of the function $v(x_0, t_0, x, t)$, and $x(t)$ is a solution of equations (1.3). Since the derivatives dx_i/dt are uniformly bounded in the region $H_0(2\gamma)$, inequality (4.14) shows that we can decrease the numbers α_m, τ_m until the inequalities

(4.15) $\dfrac{dv(x_0, t_0, x(t), t)}{dt} > d_m$

are satisfied for $|| x - x_0 ||_2 < \alpha_m$, $-\tau_m + t_0 < t < t_0 + \tau_m$. It will be convenient to assume the validity of this auxiliary inequality in the sequel.

Since the set F_m is compact, we can take a finite set of points $x_0^{(l)} \in F_m$ $(l = 1, \cdots, N_m)$ so that the system of neighborhoods $|| x - x_0^{(l)} ||_2 < \alpha_m$ $(l = 1, \cdots, N_m)$ covers F_m. Set $t_0^{(s)} = s\tau_m/2$ $(s = 0, 1, \cdots)$. To the sequence of pairs $(x_0^{(l)}, t_0^{(s)})$ there corresponds a sequence of functions $v(x_0^{(l)}, t_0^{(s)}, x, t)$ constructed by the rule given above.

We define the function

(4.16) $v_m(x, t) = \sum\limits_{s=0}^{\infty} \sum\limits_{l=1}^{N_m} v(x_0^{(l)}, t_0^{(s)}, x, t)$,

where the summation is extended over all values of $l = 1, \cdots, N_m, s = 0, 1, \cdots$. It can be seen from the above that the function v_m possesses the following properties: $v_m(x, t) = 0$ in the region $R(N(m) + 1)$; the function v_m has continuous and bounded first-order partial derivatives in all its arguments at every point of the region G. Indeed, if $t > 0$ is arbitrary, only a finite number of terms in the right member of (4.16) differ from 0 in the neighborhood of an arbitrary point x of G. Moreover, this number is a bounded constant, independent of the point x. The derivative dv_m/dt has a definite sign in the region F_m, since an arbitrary point x of F_m is contained in at most one neighborhood (4.15); the respective terms in (4.16) satisfy inequality (4.15). Now suppose the inequalities

$$(4.17) \qquad \left|\frac{\partial v_m}{\partial x_i}\right| < P_m, \quad \left|\frac{\partial v_m}{\partial t}\right| < P_m, \quad |v_m| < P_m$$

are satisfied. We plan to show that the function

$$(4.18) \qquad v(x_1, \cdots, x_n, t) = \sum_{m=k}^{\infty} \frac{1}{P_m 2^m} v_m(x_1, \cdots, x_n, t)$$

satisfies all the conditions of the theorem. First, the series in the right member of (4.18) and the series

$$\frac{\partial v}{\partial x_i} = \sum_{m=k}^{\infty} \frac{1}{P_m 2^m} \frac{\partial v_m}{\partial x_i}, \quad \frac{\partial v}{\partial t} = \sum_{m=k}^{\infty} \frac{1}{P_m 2^m} \frac{\partial v_m}{\partial t}$$

converge uniformly and absolutely in the region G, by inequality (4.16). Thus the function v exists and is continuously differentiable in G. Next, the fact that v admits an infinitely small upper bound, and the fact that dv/dt is positive-definite, follow from the properties of the function v_m already established. Thus the first assertion of the theorem is proved.

We now prove the second assertion of the theorem. If the functions $X_i(x_1, \cdots, x_n, t)$ are uniformly continuous in the time t in the region G for $0 < t < \infty$, remark (b) on p. 15 shows that the function $V = v(x_0, t_0, x, t)$ has continuous partial derivatives of all orders in every argument, and that these derivatives are uniformly bounded. (It was for this reason that X_i were required to be uniformly continuous in t.) It follows further that the function v_m also possesses continuous partial derivatives of every order in all arguments, and satisfies the inequality

$$(4.19) \qquad \left|\frac{\partial^l v_m}{\partial x_1^{l_1} \cdots \partial x_n^{l_n} \partial t^{l_{n+1}}}\right| < P_m^{(l)} \qquad [P_m^{(l)} = \text{const.}; l = 0, 1, \cdots].$$

We turn our attention now to the second assertion of the theorem. Suppose that the functions X_i are uniformly continuous in the time, $0 < t < \infty$, in the region Γ. Redefine the numbers P_m in equation (4.18) by the relations

$$(4.20) \qquad P_m = \max P_\mu^{(l)} \quad \text{for} \quad \mu, l = 1, \cdots, m.$$

Then the function v (4.18) is defined in the region G and has continuous and bounded partial derivatives of every order in all its arguments. Indeed, by relation (4.19), with P_m chosen as in (4.20), the series in the right member of (4.18) and all the series for the partial derivatives of every order do converge uniformly and absolutely in the region G. This proves the second assertion of the theorem.

We turn now to the third assertion. Suppose that $X_i(x, t)$ are periodic functions of the time t, and all have the same period ϑ. Formula (4.16) can be summed with respect to s for $-\infty < s < \infty$; in this it can be supposed that the number τ_m in formula (4.15) is a divisor of the period ϑ. Then the function $v_m(x, t)$ not only enjoys the properties adumbrated above, but also is a periodic function of the time t, with period ϑ; i.e., the function $v(x, t)$ of (4.18) is also a periodic function of the time t with period ϑ.

If the functions X_i in the right member of (1.3) are independent of the time t, we take $\vartheta > 0$ to be an arbitrary positive number and the numbers τ_m to be divisors of ϑ, in defining v_m. This tentative function v_m is denoted by $v_m^{(0)}(x_1, \cdots, x_n, t)$.

In general, we construct the function $(1/l)v_m^{(l)}(x_1, \cdots, x_n, t)$ $(l = 1, 2, \cdots)$ in the same way we constructed $v_m^{(0)}$, but dividing τ_m by l. All the functions of the sequence so defined are periodic, and the period of the lth one is ϑ/l. It can be shown that the functions $v_m^{(l)}/l$ are uniformly bounded in G as l increases; and that the functions $(1/l)[\partial^\mu v_m^{(l)}/\partial x_1^{\mu_1} \cdots \partial x_n^{\mu_n}\partial t^{\mu_{n+1}}]$ are likewise uniformly bounded (with a different bound for each partial derivative). This is so because if, in a neighborhood of every point x of G, the number of nonzero terms $v(x_0, t_0, x, t)$ in the formula for v_m is bounded by some constant N, the number of nonzero terms in a neighborhood of the same point $x \in G$ in the formula for $v_m(x, t)$ is no greater than Nl. The lth function is, however, $v_m^{(l)}/l$ (i.e., has been provided with the factor $1/l$). The assertion that the functions $v_m^{(l)}/l$ and all the partial derivatives of this function are bounded is proved.

Next, we observe that every function $v_m^{(l)}/l$ has a derivative $(1/l)dv_m^{(l)}/dt$ that is positive-definite in the region F_m; in fact the inequality

$$(4.21) \qquad \frac{1}{l} \frac{dv_m^{(l)}}{dt} > d_m$$

is satisfied, independent of l for some positive constant $d_m > 0$. This is true because if x is an arbitrary point $x \in F_m$, the number of terms $v(x_0, t_0, x, t)$ in (4.16) for which inequality (4.15) is satisfied in the formula for $v_m^{(l)}$ is bounded from below by the number l. This shows that inequality (4.21) is valid. The functions $(1/l)v_m^{(l)}$ and all the functions

$$\frac{1}{l} \frac{\partial^\mu v_m^{(l)}}{\partial x_1^{\mu_1} \cdots \partial x_n^{\mu_n}\partial t^{\mu_{n+1}}}$$

form a family of functions that are uniformly bounded and uniformly continuous. Thus by use of the diagonal process, we can construct a subsequence $v_m^{(l_\nu)}(x, t)$ so that all the sequences

$$v_m^{(l)} \quad \text{and} \quad \frac{\partial^\mu v_m^{(l_\nu)}}{\partial x_1^{\mu_1} \cdots \partial x_n^{\mu_n}\partial t^{\mu_{n+1}}}$$

converge to the respective functions

$$v_m \quad \text{and} \quad \frac{\partial^\mu v_m}{\partial x_1^{\mu_1} \cdots \partial x_n^{\mu_n}\partial t^{\mu_{n+1}}} \, .$$

The function v_m so obtained is a stationary function; that is, it is independent of the time t. Moreover, the function has a derivative satisfying

$dv_m/dt > d_m$ for $x \in F_m$, by inequality (4.21). The function (4.18), with P_m defined by (4.21) and (4.20), is also a stationary function, and satisfies all the conditions of the theorem. The theorem is completely proved.

The statement of the theorem just derived can be made stronger; the conditions imposed on the functions X_i are stronger than those needed to derive the conclusions above. Something weaker than a Lipschitz condition will do. (See [96] and [129].) It is also possible to give conditions under which a function v will exist with a derivative dv/dt of definite sign in the entire domain G of definition of the right member of (1.3); in particular, in the whole space $-\infty < X_i < \infty$. For asymptotic stability, we shall give such conditions in section 5. The most general results seem not to have enough practical interest to be stated here, the case of instability being particularly tedious.

The following theorem is an obvious consequence of theorems 4.2 and 4.3 and lemma 4.1.

THEOREM 4.4. *Let the functions X_i in the right member of equations (1.3) be periodic functions of the time t, all with the same period ϑ (or let them all be independent of the time).*

The condition that every bounded region H_0, $\bar{H}_0 \subset G$, have the property that no complete trajectory $x(x_0, t_0, t)$, $-\infty < t < \infty$ (except the trivial trajectory $x=0$) be contained in H_0 is a necessary and sufficient condition for the existence, in every such region, of a function $v(x_1, \cdots, x_n, t)$ with the following properties.

(i) *v is periodic, with period ϑ (or independent of the time t);*

(ii) *v admits an infinitely small upper bound;*

(iii) *the total derivative dv/dt, evaluated along (1.3), has a definite sign in the region H_0.*

In conclusion, we call attention to the following facts. Suppose we know some finite number of segments of the trajectories of (1.3) to within ε, where ε is the number in inequalities (3.6). If T^*, m are arbitrary positive constants, we can construct a function $v(x_1, \cdots, x_n, t)$ by the method of theorem 4.1 (see also lemmas 3.1 and 3.2) that will have a derivative of definite sign in the region $H_0 \backslash R(m)$ for $0 < t < T^*$. In this region, the function v is given in closed form, i.e., as a finite sum. Thus, if we know that the region Γ enjoys property A, a finite number of operations suffices to construct a function v in a given region $H_0 \backslash R(m)$, $0 < t < T^*$. If the functions X_i are independent of t, and it is desired to construct a function v that is also independent of t, it is also necessary to find a finite number l_0 for which $v_m^{(l_0)}$ (p. 24) is sufficiently close to $v_m(x)$, and to set $v_m(x) = v_m^{(l_0)}(x, 0)$ tentatively. To estimate an acceptable value for l_0, it is necessary to know the numbers P_m and d_m in every region $H_0 \backslash R(m)$. In a concrete problem this procedure can admittedly entail an arduous calculation.

5. Theorems on the Existence of a Function $v(x, t)$ that Satisfies the Conditions of Lyapunov's Theorem on Asymptotic Stability

In this section we shall use the basic results of sections 3 and 4 to investigate the existence of a Lyapunov function $v(x_1, \cdots, x_n, t)$ that guarantees asymptotic stability of the null solution $x = 0$ of equations (1.3). We have already pointed out that this question is considered in a number of references [13], [16], [17], [84], [89], [91], [93], [97], [125], [129], [131], [196], [197], [199]. Here we base the proof of existence of the function v on the results of section 4. We use the revised statement of Lyapunov's theorem on asymptotic stability [condition (4.5), p. 17]. The references cited consider this question of asymptotic stability only for an (arbitrary) initial perturbation in the region G. In the preceding paragraphs we showed that the function v of Lyapunov's theorem on asymptotic stability exists only if property A holds.

If the motion is uniformly asymptotically stable in the sense of the following definition, there is a characteristic property of the region that corresponds to property A in the case of ordinary stability.

DEFINITION 5.1 [125]. *The null solution $x = 0$ of equation (1.3) is called asymptotically stable uniformly with respect to the time t_0 and the coordinates of the initial perturbation x_0 in the region G_δ, if*

  (a) *the solution $x = 0$ is stable in the sense of Lyapunov (definitions 1.1 and 1.2, pp. 2-3),*

  (b) *for every positive number $\eta > 0$, there is a number $T(\eta)$ such that the inequality*

(5.1) $\| x(x_0, t_0, t) \|_2 < \eta$

is satisfied for $t \geq t_0 + T(\eta)$, independent of the initial moment of time t_0 and the coordinates of the initial perturbation x_0, $x_0 \in G_\delta$. This property of the trajectories is called property B.

The proof of Lyapunov's theorem on asymptotic stability can be modified to show that the conditions of that theorem carry with them uniform stability in the sense of definition 5.1. (For a proof, see Malkin [125].) Thus it is true that if we wish to establish the existence of a Lyapunov function v and we know the motion to be asymptotically stable, we can assume that it is uniformly asymptotically stable. (We shall show below that this is equivalent to property A.)

We emphasize that property B is not only a necessary but also a sufficient condition for the existence of a Lyapunov function v with the appropriate properties. This is the assertion of the following theorem.

THEOREM 5.1. *Let $0 \in H_0$, where H_0 is a bounded region such that $\bar{H}_0 \subset G$. If the null solution $x = 0$ of the system (1.3) is asymptotically stable uniformly with respect to the time t_0 and the initial perturbation x_0, $x_0 \in H_0$, then there is a function $v(x,t)$ that has a derivative dv/dt which is negative-definite along a trajectory of (1.3), in the region H_0. The function v is positive-definite in H_0, admits an infinitely small upper bound there, and has first-order partial derivatives in all its arguments that are uniformly bounded with respect to the time, $0 \leq t < \infty$, in H_0.*

If the functions $X_i(x_1, \cdots, x_n, t)$ are continuous uniformly with respect to the time, $0 < t < \infty$ in a region $H_1 \supset \bar{H}_0$, the function $v(x,t)$ has partial derivatives of all orders with respect to all its arguments, and these derivatives are uniformly bounded, $0 < t < \infty$ in the region H_0 (each derivative with its own bound).

PROOF. We show first that we can find a bounded region G_δ such that $\bar{H}_0 \subset G_\delta$, $\bar{G}_\delta \subset G$, and the solution $x = 0$ is also uniformly asymptotically stable with respect to initial perturbations in G_δ (i.e., has property B). (The value of the constant $T(\eta)$ need not be the same for G_δ as for H_0.)

If η_0 is a fixed positive number such that $\eta_0 < \frac{1}{2}\rho(\bar{H}_0 \backslash H_0, 0)$, and if $t_0 > 0$ is an arbitrary moment of time, we can find a bounded region Γ, $\bar{\Gamma} \subset G$ with the property that the trajectory $x(x_0, t_0, t) \in \Gamma$ for $t \geq t_0$ whenever $x_0 \in H_0$; this is definition 1.2. Let $2\gamma = \rho(\bar{\Gamma}\backslash\Gamma, \bar{G}\backslash G)$. If this distance ρ is infinite, we take an arbitrary positive number for γ. Property B of definition 5.1 yields $\| x(x_0, t_0, t_0 + T(\eta_0)) \|_2 < \eta_0$ for all $x_0 \in H_0$. Let $H_0(\gamma)$ be the set of points for which $\rho(x, \Gamma) < \gamma$. It is clear that $\bar{H}_0(\gamma) \subset G$. Let δ be defined by $2\delta = \min(\gamma, \eta_0) \exp[-nL_\gamma T(\eta_0)]$; let G_δ be the set of points x for which $\rho[H_0, x] < \delta$. By use of an argument similar to that on p. 21 (see [138, p. 23]) it can be shown that, for all values of t for which the segments $[t_0, t]$ of both trajectories $x(x_0', t_0, t)$ and $x(x_0'', t_0, t)$ lie entirely in the region H_γ, the inequality

$$(5.2) \qquad \| x(x_0'', t_0, t) - x(x_0', t_0, t) \|_2 < \| x_0'' - x_0' \|_2 \exp[nL_\gamma | t - t_0 |]$$

holds if $x_0' \subset G_\gamma$ and $x_0'' \subset G_\gamma$. If we recall how the regions $H(\gamma), G_\gamma$ were constructed, we see that for $t_0 \leq t \leq t_0 + T(\eta_0)$, the trajectory $x(x_0'', t_0, t)$ lies in the region $H(\gamma)$ for $x_0'' \subset G_\gamma$. Moreover,

$$\| x(x_0'', t_0, t_0 + T(\eta_0)) \|_2 < 2\eta_0; \quad \text{i.e.,} \quad x(x_0'', t_0, t_0 + T(\eta_0)) \in H_0 .$$

Indeed, to show that this assertion is valid, we need only turn to inequality (5.2) and choose the point x_0' so that it is in H_0 and the relation $\| x_0'' - x_0' \|_2 \leq 2\delta$ is satisfied. By definition 5.1, property B, the inequality

$$\| x(x_0, t_0, t_0 + T(\eta_0) + t) \|_2 < \eta$$

will be valid for all $t > T(\eta)$ provided the trajectory enters H_0 for $t = t_0 + T(\eta_0)$.

Let $T(\eta)$ be the constant of property B, definition 5.1, for initial perturbations x_0 in the region H_0. Set $T_\delta(\eta) = T(\eta_0) + T(\eta)$. If we take $T_\delta(\eta)$ as the constant of this definition for the region G_δ, the assertion made at the beginning of the proof is established.

We now show that property A of definition 4.1 is satisfied in the region G_δ. We take

$$t_0 \geq T_{\delta(\eta)} , \qquad x_0 \in [G_\delta \backslash R(\eta)] ,$$

where $R(\eta)$ is the region $\| x \|_2 < \eta$. The point $x(x_0, t_0, t_0 - T_\delta)$ cannot lie in the region G_δ since property B would give

$$\| x(x(x_0, t_0, t_0 - T_\delta(\eta)), t_0 - T_\delta(\eta), t_0) \|_2 = \| x_0 \|_2 < \eta ,$$

and this is a contradiction.

Thus property A is satisfied in the region G for the negative half-trajectory. On the basis of theorem 4.2, we can construct a function $v(x_1, \cdots, x_n, t)$ that is defined in the region H_0 and satisfies all the conditions of the theorem, except possibly the property of being positive-definite in the region H_0. (The function v can be taken to be the negative of the function of theorem

4.2.) To see that the function v is positive-definite in the region H_0, we note that all terms of the formulas (4.16) and (4.18) used to construct the function $v(x_0, t_0, x, t)$ in lemma 3.2 are negative. Moreover, if x_0 is arbitrary, $x_0 \in [H_0 \backslash R(m)]$, at most one term in (4.16) satisfies the inequality

$$v(x_0^{(l)}, t_0^{(s)}, x, t) < -\varDelta_m < 0 \qquad (\varDelta_m = \text{const.})$$

because of (4.15) and the choice of the numbers $t_0^{(s)}$. The theorem is proved.

We now turn to the case in which the right members of equations (1.3) are periodic functions of the time, all with the same period ϑ, or are independent of the time. In such a case, if we consider initial perturbations in a bounded closed region $\bar{H}_0$, asymptotic stability of the null solution $x = 0$ is always uniform with respect to the initial coordinate x_0. (See [125].) This assertion will be proved in Chapter 2, in which the connection between asymptotic stability and the properties of Lyapunov functions is covered in detail. If we anticipate this result, we can establish the following corollary to theorem 5.1.

COROLLARY 5.1. *Let $\bar{H}_0$ be a bounded closed region, $\bar{H}_0 \subset G$. Suppose that the null solution $x = 0$ of equations (1.3) is asymptotically stable, and that the region $\bar{H}_0$ lies in the region of attraction of the point $x = 0$. Then there is a region $H_1 \supset \bar{H}_0$ in which a function $v(x, t)$ can be constructed that is periodic in the time t with period ϑ (or a function $v(x)$ which is independent of the time), that admits an infinitely small upper bound, and that is positive-definite in the region H_1. Moreover, the derivative dv/dt along a trajectory of (1.3) is negative-definite in the region H_1. The function v has continuous partial derivatives of all orders with respect to all its arguments, and each one is uniformly bounded in the region H_1. (The bounds themselves may vary with the particular derivative.)*

This solves the problem of obtaining a valid converse to Lyapunov's theorem on asymptotic stability. It must be mentioned here that theorems 5.1 and 5.2 do not provide a converse to the *extended* Lyapunov theorem on asymptotic stability (theorem 4.1, p. 17). Such a converse would assert the existence of a function v that satisfied the conditions of this theorem in the region G_δ of initial perturbations, and then infer the existence of a function v in the region Γ that satisfied condition (4.5) as well as the other conditions of the theorem. Such a general converse theorem is not valid. The following is an easy counterexample.

Let $dx/dt = X(x, t)$ be a first-order equation with trajectories that are symmetric with respect to the time (see Fig. 2). Suppose the single trajectory $x(x_0', 0, t)$ approaches $x = H_0$ asymptotically ($H_0 > G_\delta$); suppose that for all $t_0 > 0$, the region $|x| < G_\delta$ lies in the region of attraction of the point $x = 0$. We can even think of an example in which the trajectories $x(x_0, t_0, t)$ approach 0 uniformly in x_0, t_0, in the region $t \geq 0, |x| < H_\delta < G_\delta$.

In this example we cannot construct a function $v(x, t)$ for $0 < t < \infty, |x| < H$ that is positive-definite in this region and for which the derivative dv/dt is negative-definite along the trajectories $0 < t < \infty, |x| < H$. Theorem 4.2,

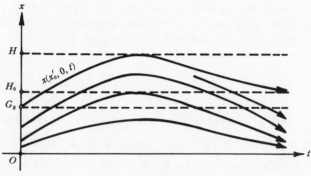

FIGURE 2

p. 19, insofar as estimates of the regions G_δ and H_0 are concerned, does not settle the problem for arbitrarily large moments of time $t_0 > 0$, but only concerns the existence of a function v for those moments at which the initial perturbation occurs (here, at $t_0 = 0$).

If the right members of equation (1.3) are independent of the time, Lyapunov's theorem on asymptotic stability, in its extended form with condition (4.5), always has a valid converse, as will be seen later.

There is a case in which Lyapunov's theorem on asymptotic stability in extended form does have a valid converse. This occurs when every infinite half-trajectory $t \geqq t_0$ remains in the region G, $x(x_0, t_0, t) \in G$ as long as it starts in G: $x_0 \in G$; the region G is the one in which asymptotic stability of the null solution $x = 0$ is assumed to occur (with respect to initial perturbations x_0). This case is naturally called "stability in the large."

In particular, if the region G is the entire space $-\infty < x_i < \infty$ $(i=1, \cdots, n)$, the problem of asymptotic stability is certainly "in the large." In a problem in which an arbitrarily large initial perturbation can occur, such a point of view is useful, but it is also useful in case there is a way of estimating beforehand the magnitude of possible initial perturbations x_0.

We next state a theorem that gives a condition sufficient for asymptotic stability in the large in the region G. The following definitions are necessary preliminaries.

DEFINITION 5.2. *The function $v(x, t)$ admits an infinitely great lower bound on the boundary of the region G (or simply, admits an infinitely great lower bound in G, or is radially unbounded in G),* [*] *if a continuous function $w(x_1, \cdots, x_n)$ exists in G such that*

$$v(x, t) \geqq w(x) \qquad \text{everywhere in } G;$$

and

$$\lim w(x) = \infty, \quad as \quad \xi(x) \to \infty,$$

where $\xi(x) = \max \{\rho[0, x], 1/\rho[\bar{G}\backslash G, x]\}$.

[*] The term *radially unbounded* is used by Hahn [66], [68].

DEFINITION 5.3. *The null solution $x = 0$ is called uniformly asymptotically stable in the large in the region G (or, property B is said to hold in the large) if for arbitrary preassigned positive $\eta > 0$ and arbitrary $H_0, \bar{H}_0 \subset G$, there are always a number $T(H_0, \eta)$ and a bounded region $H_1, \bar{H}_1 \subset G$ such that the relations*

(5.3) $x(x_0, t_0, t) \in H_1$ *for all* $t \geq t_0,$

(5.4) $\| x(x_0, t_0, t) \|_2 < \eta$ *for all* $t \geq t_0 + T(H_0, \eta),$

hold for every initial moment of time t_0 and for every given value of $x_0 \in H_0$.

A sufficient condition for asymptotic stability in the large is the following.

THEOREM 5.2. *The null solution $x = 0$ of equations* (1.3) *is asymptotically stable in the large in the region G if there exists a function $v(x, t)$ such that*

(i) $v(x, t)$ *is positive-definite in G;*

(ii) $v(x, t)$ *admits an infinitely small upper bound in G;*

(iii) $v(x, t)$ *admits an infinitely great lower bound on the boundary of G* ($v(x, t)$ *is radially unbounded in G).*

(iv)   *The derivative dv/dt along a trajectory of* (1.3) *is negative-definite in G.*

This theorem was established in [15]–[17] for the case in which G is the entire space $-\infty < x_i < \infty$; the general result was established in [97] and [131]. The theorem is incorrect without assumption (iii); a counterexample appears in [16].

Lyapunov's methods are sufficient to prove the theorem; we give the proof here to emphasize the uniform character of the resulting stability.

PROOF. Let $\bar{H}_0$ be a bounded closed region, $\bar{H}_0 \subset G$. We write $M(H_0) = \sup v, x \in H_0, 0 < t < \infty$. Clearly $M(H_0) < \infty$, since $v(x, t)$ admits an infinitely small upper bound in G. Since the function v admits an infinitely great lower bound on the boundary of G, there is a bounded region H_1 such that $\bar{H}_0 \subset H_1$, $\bar{H}_1 \subset G$, $v(x, t) > M(H_0)$ for $x \in G \backslash H_1$.

Let $x_0 \in H_0$, and consider the trajectory $x(x_0, t_0, t)$. Since $dv/dt < 0$ along this trajectory, it follows that $x(x_0, t_0, t) \in H_1$ for $t \geq t_0$. There is a constant $\alpha = \alpha(\eta, H_1)$ such that the inequality $dv/dt < -\alpha(\eta, H_1)$ will be satisfied for the entire time during which the trajectory $x(x_0, t_0, t)$ lies in H_1 and outside the neighborhood $\| x \|_2 < \eta$. Consequently, the inequality

$$v(x(x_0, t_0, t), t) < v(x_0, t_0) - \int_{t_0}^{t} \alpha \, dt = v(x_0, t_0) - \alpha(t - t_0)$$

will be satisfied.

Thus the length of time during which a segment of the trajectory $x(x_0, t_0, t)$ lies entirely outside the neighborhood $\| x \|_2 < \eta$ is bounded by the number $T = M(H_0)/\alpha$. This shows that for some $t > t_0 + T$, the trajectory enters the neighborhood $\| x \|_2 < \eta$, by virtue of an argument similar to one of Lyapunov [114, p. 261]; see p. 9. Thus the asymptotic stability of the null solution $x = 0$ in the large in the region G is proved. It is worth remarking that the region H_1 and the number T do not depend on the choice of the point $x_0 \in H_0$; they depend only on the region H_0.

We now show that the condition of uniform asymptotic stability in the large in the region G is in its turn a sufficient condition for the existence of a function $v(x, t)$ that satisfies the conditions of theorem 5.2.

THEOREM 5.3. *If the null solution* $x = 0$ *of equations* (1.3) *is uniformly asymptotically stable in the large in the region* G, *there is a function* $v(x, t)$ *such that*

(i) v *is positive-definite in* G;

(ii) v *admits an infinitely small upper bound in* G;

(iii) v *admits an infinitely great lower bound on the boundary of* G;

(iv) *the derivative* dv/dt *is negative-definite along a trajectory of the system* (1.3);

(v) *the partial derivatives* $\partial v/\partial x_i$, $\partial v/\partial t$ $(i = 1, \cdots, n)$ *are continuous and bounded in every bounded subregion* $\bar{H}_0$, $\bar{H}_0 \subset G$, *uniformly with respect to the time*, $0 < t < \infty$.

If the functions X_i *in the right member of equation* (1.3) *are uniformly continuous in the time* t *in every bounded region* H_0 $(\bar{H}_0 \subset G)$, *the function* v *has continuous partial derivatives of all orders with respect to all its arguments; these are uniformly bounded in time in every region* H_0, $0 < t < \infty$. (*Each derivative has its own constant bound.*)

If the functions X_i *are periodic functions of the time, all with the same period* ϑ (*or if they are independent of the time*), *there is a function* v *that is a periodic function of the time with period* ϑ (*or is independent of the time*), *and has properties* (i), (ii), (iii), (iv), *and* (v).

PROOF. Let H_0 be some bounded region, $\bar{H}_0 \subset G$. By theorems 4.2 and 5.1, there is a function $v_0(x_1, \cdots, x_n, t)$, which is defined in the region G and has the following properties.

(i) Let $3\gamma = \rho[\bar{G}\backslash G, H_0]$, $H_0(2\gamma) = \{x \mid \rho[H_0, x] < 2\gamma\}$. Outside the region $H_0(2\gamma)$ the function v_0 is identically 0.

(ii) v_0 is positive-definite in the region H_0.

(iii) v_0 admits an infinitely small upper bound.

(iv) The value of the derivative dv/dt along the equations (1.3) of perturbed motion is negative-definite in the region H_0.

(v) In the region $H_0(2\gamma)\backslash\bar{H}_0$, the function v_0 has a positive derivative dv_0/dt, which satisfies the inequality

$$(5.5) \qquad \left|\frac{dv_0}{dt}\right| < N_0 \qquad (N_0 = \text{const.}) .$$

Now let $\{H_m\}$, $m = 0, \pm 1, \pm 2, \cdots$, be a sequence of regions with the following properties:

(i) $\bar{H}_m \subset H_{m-1}$;

(ii) $\bigcup_m H_m = G$;

(iii) if $\varepsilon > 0$ is an arbitrary positive number, there is a number N_ε such that $H_m \subset R(\varepsilon) \equiv \{x \mid \|x\|_2 < \varepsilon\}$ for $m \geq N_\varepsilon$;

(iv) $H_{-1} = H_0(2\gamma)$.

It is clear that such a sequence always exists.

Set $F_m = H_{m-1} \setminus H_m$. Using the construction in the proof of theorem 4.2, we can define functions v_m, $m = -1, -2, \cdots$, which satisfy the following conditions:

(5.6) $\qquad \dfrac{dv_m}{dt} < -d_m < 0 \qquad$ in the region F_m,

(5.7) $\qquad \dfrac{dv_m}{dt} \leqq 0 \qquad$ in the region $\bar{H}_m$,

(5.8) $\qquad v_m = 0$ in $H_{\mu(m)}$; $\quad v_m = 0$ in the region $G \setminus \bar{H}_{m-1}$,

(5.9) $\qquad \left| \dfrac{dv_m}{dt} \right| < M_m \qquad$ in the region F_{m-1},

(5.10) $\qquad v_m > Q_m > 0 \qquad$ in the region F_m.

It is important to note that regions $H_{\mu(m)}$ exist where $v_m \equiv 0$. Moreover, as m decreases, the index $\mu(m)$ of such a region $H_{\mu(m)}$ also decreases indefinitely. This assertion can be established as follows. First we apply lemma 3.2 to conclude that the terms $v(x_0, t_0, x, t)$, which are summed in formula (4.16) to construct the function v_m, can be taken to be 0 in the region $H_{\nu+1}$, if the corresponding segment of the trajectory $x(x_0, t_0, t)$, $-\theta + t_0 \leqq t \leqq t_0$, along which the function $v(x_0, t_0, x, t)$ was constructed, lies outside the region H_ν. Next we observe that definition 5.3, property B, shows that if the point x_0 lies outside the region H_m, then every segment of the trajectory $x(x_0, t_0, t)$ will lie outside the region $H_{\nu(m)}$ for $t < 0$. Finally, we note that as m decreases, $\nu(m)$ also decreases. The assertion made above is thus verified by the choice $\mu(m) = \nu(m) + 1$. We now multiply the function v_m by a positive constant, if necessary, to obtain the results

$$v_m > -m \quad \text{in the region} \quad F_m \quad \text{for} \quad m = -1, -2, \cdots$$

and

$$\frac{dv_m}{dt} < -2 \sum_{k=m+1}^{0} \left| \frac{dv_k}{dt} \right| \quad \text{in the region} \quad F_m \quad \text{for} \quad m = -1, -2, \cdots.$$

It is now easy to verify that the function

$$v(x, t) = \sum_{m=-\infty}^{m=0} v_m(x, t)$$

satisfies all the conditions of the theorem. The details are not included. Only obvious properties of the functions v_m are involved.

We showed by example (Fig. 2, p. 29) that the extended Lyapunov theorem has no valid converse if the region of attraction H_0 is defined by something like relation (4.5), if we require that the function $v(x, t)$ be defined in the region Γ for all $t \in (0, \infty)$ and be independent of the time. However, this theorem can be modified in such a way that the modified theorem admits a converse in all cases, and at the same time the statement of the theorem includes conditions similar to inequality (4.5) for estimating the region of attraction G_δ. The modification consists in permitting .he region

$\varGamma$ in which the function $v(x, t)$ is defined, and also the region H_0, on the boundary of which inf v exceeds (sup v in G_δ), to depend on the time t. The corresponding modification of theorem 4.1, p. 17, is as follows. The null solution $x = 0$ is asymptotically stable, and the region G_δ lies in the region of attraction of the point $x = 0$, provided it is possible to construct a function $v(x, t)$ with the following properties.

(i) The function $v(x, t)$ is defined in a region $G(t) \subset \varGamma$ for every t, $0 < t < \infty$.

(ii) The function $v(x, t)$ is positive-definite in $G(t)$.

(iii) The function $v(x, t)$ admits an infinitely small upper bound in $G(t)$.

(iv) The total derivative dv/dt along a trajectory of (1.3) is negative-definite in $G(t)$.

(v) For every t, $0 < t < \infty$, there is a connected region $G_0(t)$ which decreases continuously with time, and is such that $\bar{G}_0(t) \subset G(t)$;

(5.11) $[\sup v(x, t) \text{ in } G_\delta] < [\inf v(x, t) \text{ in } \bar{G}_0(t) \backslash G_0(t)]$.

To prove this assertion, let $x_0 \in G_\delta$, and consider the trajectory $x(x_0, t_0, t)$. From inequality (5.11) we see that the trajectory in question cannot leave the region $G_0(t)$ for $t \geq t_0$, since the derivative dv/dt is negative for all $\tau \leq t$ in the region $G(\tau)$, which contains $G_0(\tau)$. Thus for $t \geq t_0$, we have $x(x_0, t_0, t) \in G(t) \subset \varGamma$. Moreover, the derivative dv/dt is negative-definite in the region $G(t)$, so the proof can be completed in the same manner as in Lyapunov's book [114, p. 261]; see p. 9. We remark further that the estimates obtained during the proof show that the solution $x = 0$ is asymptotically stable in the sense of definition 5.1 (property B obtains), with respect to initial perturbations x_0 in the region G_δ.

Conversely we can show that if the solution $x = 0$ is uniformly asymptotically stable in the sense of definition 5.1, i.e., uniformly with respect to x_0 and the time t_0 ($x_0 \in G_\delta$), property B, then there is a function $v(x, t)$ that satisfies all the conditions given above. This is the assertion of the following theorem.

THEOREM 5.4. *If the null solution $x = 0$ of equation* (1.3) *is uniformly asymptotically stable in the sense of definition* 5.1 (*property* B), *and if G_δ lies in the region of attraction, there is a function $v(x, t)$ defined in a region $G(t)$ for all $t > 0$ ($\bar{G}_\delta \subset G(t) \subset \varGamma$); v is positive-definite in $G(t)$; v admits an infinitely small lower bound in $G(t)$; and the total derivative dv/dt along a trajectory of* (1.3) *is a negative-definite function in $G(t)$. Moreover, for each t, $0 < t < \infty$, there is a uniformly bounded region $G_0(t)$, $\bar{G}_0(t) \subset G(t)$, for which inequality* (5.11) *is satisfied.*

PROOF. Here we prove this assertion only when the functions X_i satisfy somewhat more stringent conditions than those made above. We suppose that $X_i(x, t)$ in the right members of equations (1.3) have continuous partial derivatives $\partial X_i/\partial x_j$ ($i = 1, \cdots, n; j = 1, \cdots, n$), which are uniformly bounded in the region $\varGamma$ with respect to the time, i.e.,

(5.12) $\left| \dfrac{\partial X_i}{\partial x_j} \right| < L$, $x \in \varGamma$, $0 < t < \infty$.

As shown in the beginning of the proof of theorem 5.1, there is a region H_1 such that $\bar{G}_\delta \subset H_1$, $\bar{H}_1 \subset \Gamma'$, and the solution $x = 0$ is asymptotically stable in the sense of definition 5.1, i.e., uniformly with respect to the initial perturbation x_0 in the region H_1 (property B). Let $G(t)$ be a region that contains the points $x(x_0, t_0, t)$, where $x_0 \in H_1$ and t_0 satisfies $0 < t_0 < t$. Let $\bar{G}_\beta(t)$ be the closed set of points $x(x_0, t_0, t)$ for $0 \leqq t_0 \leqq t$, $x_0 \in \bar{G}_\delta$. We construct the function $v_1(x, t)$ in the region $G(t)$ as follows. Let $\varphi(\tau)$ be a continuous function of τ that decreases monotonically to 0 for $\tau \to \infty$, chosen to satisfy

$$(5.13) \qquad \| x(x_0, t_0, \tau) \|_2^2 \leqq \varphi(\tau - t_0) \quad \text{for all} \quad \tau > t_0, \quad x_0 \in G(t_0).$$

The fact that the null solution is uniformly asymptotically stable is enough to guarantee the existence of such a function $\varphi(\tau)$; for details, see for example [124, p. 306]. At the same time (same reference) it can be shown that there is a monotonically increasing function $V(\varphi)$ which has a monotonically increasing derivative, and for which the conditions

$$(5.14) \qquad \int_0^\infty V(\varphi(\tau)) d\tau = N_1 < \infty,$$

$$(5.15) \qquad \int_0^\infty V'(\varphi(\tau) e^n)^{L\tau} d\tau = N_2 < \infty$$

are also satisfied. The function $v(x_1, \cdots, x_n, t)$ defined by the formula

$$(5.16) \qquad v(x_{10}, \cdots, x_{n0}, t_0) = \int_{t_0}^\infty V(\| x(x_0, t_0, \tau) \|_2^2) \, dt$$

is continuously differentiable in the region $G(t)$ in each of the arguments $x_1, \cdots, x_n, t$. It has a negative-definite derivative dv/dt in this region; the function $v(x, t)$ is positive-definite in $G(t)$. Formula (5.16) and conditions (5.14) and (5.15) are enough to establish the truth of this formula. See for example [124, pp. 300–307]. Set $M = \sup v(x, t)$ for $x \in \bar{G}_\delta, 0 < t < \infty$. We now construct a continuous differentiable function $f(x_1, \cdots, x_n, t)$ defined in the region $G(t)$ for $0 < t < \infty$, which satisfies the conditions

$$(5.17) \qquad f(x, t) = \| x \|_2^2 \quad \text{in the region } G_\beta(t),$$

$$(5.18) \qquad V(f(x, t)) > Q \quad \text{in the region} \quad \rho[x, G_\beta(t)] > \frac{1}{4} \rho[\bar{G}(t) \backslash G(t), G_\beta(t)],$$

where

$$Q = 4MnL \frac{1}{\alpha} \max (\| x \|_2^2 \quad \text{for} \quad x \in \bar{\Gamma} \backslash \Gamma),$$

$$\alpha = \min [\bar{G}(t) \backslash G(t), G_\beta(t)] \quad \text{for} \quad 0 < t < \infty.$$

The fact that such a function $f(x, t)$ exists can be proved by defining a function that is possibly discontinuous on the boundary of $G_\beta(t)$, and then smoothing out this discontinuity. Without going into details, we note that under the conditions given it is possible to construct a function $f(x, t)$ that has derivatives $\partial f / \partial x_i (i = 1, \cdots, n)$ everywhere in the region Γ', and that these derivatives will be bounded uniformly with respect to t, $0 < t < \infty$.

The function $v(x, t)$ defined by the formula

$$(5.19) \qquad v(x_{10}, \cdots, x_{n0}, t_0) = \int_0^\infty V(f(x(x_0, t_0, \tau), \tau)) \, d\tau$$

will satisfy all the conditions of the theorem. We verify only condition (5.11). By (5.17), we see that $v = v_1$ in the region G_δ; we plan to establish the inequality

$$(5.20) \qquad v(x, t) > M \quad \text{for} \quad \rho[x, G_\beta(t)] > \frac{\alpha}{2}, \, x \in G(t) .$$

When this is established, we shall be able to choose for $G_0(t)$ the set of points defined by $\rho[x, G_\beta(t)] < \alpha/2$, and condition (5.11) will be verified. Let x_0 be a point in the region (5.20). If $\vartheta = \alpha/(4nLR)$, $R = \max(\|x\|_2, x \in \bar{\Gamma}\backslash\Gamma)$, the estimate

$$v(x_0, t_0) = \int_{t_0}^\infty V(f(x(x_0, t_0, \tau), \tau)) \, dt > \int_t^{t+\vartheta} V(f) \, d\tau$$

is valid.

For $t \geq \vartheta = \alpha/(4nLR)$, the trajectory $x(x_0, t_0, t)$ remains in the region $\rho[x, G_\beta(t)] > \alpha/4$, (5.18) is satisfied, and the point x_0 satisfies the inequality $v(x_0, t_0) > \vartheta Q > M$; thus inequality (5.20) follows.

We note further that the function $v(x, t)$ has uniformly bounded partial derivatives $\partial v/\partial x_i$ in the region $G(t)$ of its definition, because, if $x \in G(t)$,

$$(5.21) \qquad v(x_0, t_0) = \int_{t_0}^\infty V(f) \, dt = \int_{t_0}^{t_0+\theta} V(f) \, d\tau + v(x(t_0 + \theta), t_0 + \theta) ,$$

where $x(t_0 + \theta) \in G_\delta$.

Since the solution $x = 0$ is asymptotically stable uniformly with respect to the initial point in the region H_1 (hence also in $G(t)$), the numbers θ defined above are uniformly bounded.

The first term of (5.21) is an integral that depends on the coordinates x_{i0} as parameters. By the theorem of [39, pp. 23–30], the solutions $x_i(x_0, t_0, t)$ have continuous partial derivatives $\partial x_i/\partial x_{j0}$, which are uniformly bounded for $t_0 \leq t \leq t_0 + \theta$. Moreover, the functions V, f, $x_i(x_0, t_0, t)$ are assumed to be continuous and differentiable in their arguments. Since an integral can be differentiated under the integral sign with respect to parameters [186a, pp. 67, 74], the integral term of (5.21) has uniformly bounded partial derivatives with respect to x_{j0}. The second term of (5.21) also has uniformly bounded partial derivatives $\partial v_1/\partial x_{j0}$ for $x_0 \in G_\delta$. (See p. 33.) Thus the function $v(x, t)$ is differentiable in the region $G(t)$. The truth of our assertion is established.

6. Theorems on the Existence of a Function $v(x, t)$ that Satisfies the Conditions of Lyapunov's First Theorem on Instability

In this section we shall use the results of sections 3 and 4 to investigate the existence of a Lyapunov function $v(x, t)$ that satisfies the conditions of his first theorem on instability. This theorem appears on p. 9; see [114, theorem II, p. 262]. Let us first consider the general case of unstable motion, i.e., the case in which the functions $X_i(x, t)$ in the right member of equation (1.3) depend explicitly on the time. The function $v(x, t)$ in the theorem of Lyapunov that applies to this case has a derivative, dv/dt, of definite sign along a trajectory of (1.3) and admits an infinitely small upper bound. By theorem 4.3, this is only possible if there is a region which contains the unperturbed solution $x = 0$, and which has property A of definition 4.1. That is, when theorem II [114, p. 262] applies, not only is the null solution $x = 0$ unstable, but some neighborhood of the point $x = 0$ enjoys property A. Thus, when the null solution $x = 0$ is unstable, the existence or non-existence of the Lyapunov function v in his second theorem depends only on the satisfaction of property A. The following theorem asserts the existence of a Lyapunov function v.

THEOREM 6.1. *If the null solution $x = 0$ of equation (1.3) is unstable in a region* Γ *that enjoys property A (p. 19), then, if H_0 is a bounded region, $\bar{H}_0 \subset \Gamma$, a Lyapunov function $v(x, t)$ must exist that satisfies the conditions of Lyapunov's first theorem on instability. This function v, then, has the following properties.*

(i)   *v admits an infinitely small upper bound.*

(ii) *The derivative dv/dt has a definite sign in the region H_0.*

(iii) *For every positive $t_0 > 0$, there is a sequence of points $x_0^{(k)}$ such that $\lim_{k \to \infty} x_{i0}(k) = 0$ $(i = 1, \cdots, n)$, and sign $v(x_0^{(k)}, t_0) =$ sign dv/dt.*

(iv) *The function $v(x, t)$ has continuous partial derivatives $\partial v/\partial x_i$, $\partial v/\partial t$ $(i = 1, \cdots, n)$, which are uniformly bounded in the compound region $x \in H_0$, $0 \leq t < \infty$.*

If the functions X_i are uniformly continuous in the time for $x \in H_0$, $0 < t < \infty$, the function v has continuous partial derivatives of every order with respect to all arguments, and these are uniformly bounded for $0 \leq t < \infty$ (each derivative having its own bound).

PROOF. By theorem 4.2, a function can be defined in the region H_0 that has all the properties required by theorem 6.1 except possibly property (iii).

Let $v_1(x, t)$ denote the function whose existence is assured by theorem 4.2. Let H_1 be a bounded region, $\bar{H}_0 \subset \bar{H}_1 \subset \Gamma$. Let $x_0^{(k)} \in H_0$ $(k = 1, 2, \cdots)$ be a sequence of points that converges on the point $x = 0$, and is such that the trajectories $x(x_0^{(k)}, 0, t)$ enter the region H_1 with increasing t, and let this penetration occur for the first time for $t = t_k > 0$. By lemma 3.1, it is possible to construct a function $v^{(k)}(x, t)$ that is positive in a neighborhood of the segment $x(x_0^{(k)}, 0, t)$ $(0 < t < t_k)$, whose derivative is positive in this neighborhood. Thus, the inequality $dv^{(k)}/dt \geq 0$ is satisfied in the region H_0. Further, the functions $v^{(k)}$ have continuous partial derivatives of arbitrary order in all arguments in the region $x \in H_0$, $t > 0$. Let N_k be a sequence of numbers

---

* In other words, if $\bar{H}_0 \in \Gamma$ is a bounded closed subregion of Γ, there must be a sequence of points $x_0^{(k)}$ converging to the point $x = 0$ so that the trajectory $x(x_0^{(k)}, t_0, t)$ leaves the region H_0 as the time t increases. Moreover, it must be possible to define such a sequence for every fixed value of $t_0 > 0$.

subject to the inequalities

(6.1) $$|v^{(k)}| < N_k, \qquad \left| \frac{\partial^\mu v^{(m)}}{\partial x_1^{\mu_1} \cdots \partial x_n^{\mu_n} \partial t^{\mu_{n+1}}} \right| < N_k$$

for $0 \leq t < \infty$ and for $m = 1, \cdots, k$. The existence of such a sequence of numbers follows from the fact that each of the functions $v^{(k)}$ can be chosen to have values which differ from 0 at least for the interval $t < t_k + 2\tau_k$, where t_k, τ_k depend on $x_0^{(k)}$. We construct the function

(6.2) $$v_2(x, t) = \sum_{k=1}^{\infty} \frac{1}{2^k N_k} v^{(k)}(x, t) .$$

The function v_2 is defined in the region $x \in H_0$, $0 < t < \infty$, and in this re-- gion v_2 has continuous partial derivatives in all its arguments. These are all uniformly bounded (each with its own constant bound), since series (6.2) and the series obtained by differentiating (6.2) converge uniformly and absolutely in the region $x \in H_0, 0 < t < \infty$. The properties of the functions $v^{(k)}$ specified above thus show that the relation

(6.3) $$v_2(x_0^{(k)}, 0) = \alpha_k > 0 \quad (k = 1, 2, \cdots)$$

is satisfied.

By theorem 4.2, the function v_1 has uniformly bounded partial derivatives $\partial v_1(x, t)/\partial x_i$. Thus the function v_1 satisfies the inequality

(6.4) $$|v_1(x, t)| < M \|x\|_2 \quad (x \in H_0, M = \text{const.}) .$$

Let $f(r)$ be a function that has continuous derivatives of all orders, and such that

(6.5) $$f(M \| x_0^{(k)} \|) < \frac{1}{2} \alpha_k ,$$

and that $f'(r) > 0$ for $r \neq 0$. (It is easy to show how to construct such a function.)

We now construct the function $v = f(v_1) + v_2$, and show that v satisfies all the conditions of the theorem. Without expounding all details, we show in particular that, if positive $t_0 > 0$ is arbitrary, we can construct a sequence of points for which $v > 0$, and this sequence must converge to the point $x = 0$. If $x_0^{(k)}$ is such a sequence, there must be a sufficiently small neighborhood $u_k(x_0^{(k)})$ of $x_0^{(k)}$ in which the inequality $v > 0$ is satisfied for $t = 0$, by inequalities (6.3), (6.4), and (6.5). The inequality $v > 0$ will therefore continue to be satisfied for all values $t > 0$ for which the trajectory $x(x_0, 0, t)$ remains in the region H_0, since the derivative dv/dt is positive-definite along this trajectory, assuming that x_0 lies in the neighborhood u_k. Therefore, the inequality $v(x(x_0, 0, t_0), t_0) > 0$ holds for all sufficiently large k, for all $x_0 \in u_k$, $t_0 > 0$. Since the solution $x(x_0, t_0, t)$ depends continuously on the initial conditions, it follows that $\lim x(x_0, 0, t_0) = 0$ (as $x_0 \to 0$), so that the set of points $x(x_0, 0, t_0)$ with $x_0 \in u_k$ has the origin $x = 0$ for a limit. Moreover, the inequality $v > 0$ is satisfied at every point of this set. The remaining assertions of the theorem are established easily; for example, the smoothness of the function $v(x, t)$ follows immediately from the corresponding property of the functions v_1, v_2. See remark (b) after lemma 3.2 (p. 15) and also theorem 4.2.

Now we turn to the case in which the right members X_i of equations (1.3) are periodic functions of the time, all with the same period ϑ, or do not depend explicitly on t. In these cases we have already seen that property A is equivalent to the requirement that no entire trajectory (except $x = 0$) of the system (1.3) lie in a bounded region H_0, $H_0 \subset \Gamma$ (lemma 4.1, p. 20). The following theorem summarizes the situation for instability in this case.

THEOREM 6.2. *Let the functions* $X_i(x, t)$ *of equations* (1.3) *be periodic functions of the time* t, *all with period* ϑ, *or let them not depend explicitly on* t. *Let the null solution* $x = 0$ *of* (1.3) *be unstable in the region* Γ. *(See footnote, p. 36.) Suppose further that no bounded region* H_0 ($H_0 \subset \Gamma$) *contains an entire trajectory (not* $x = 0$) *of the system* (1.3). *Let* H_0 *be such a region (i.e.,* $H_0 \subset \Gamma$). *Then a function* $v(x, t)$ *exists that satisfies the conditions of Lyapunov's first theorem on instability in this region, i.e.,*

(i) $v(x, t)$ *is defined in* H_0, *and admits an infinitely small upper bound there.*

(ii) dv/dt *has a definite sign in* H_0.

(iii) *For every* $t > 0$, *one can find a sequence of points* $x_0^{(k)}$ *that converges to* 0 ($k \to \infty$) *and is such that sign* $v(x_0^{(k)}, t) = $ *sign* dv/dt.

(iv) *The function* $v(x, t)$ *is periodic in* t, *with period* ϑ *(or does not depend explicitly on* t).

(v) *The function* $v(x, t)$ *has continuous derivatives of all orders with respect to all arguments, for* $x \in H_0$.

PROOF. By theorem 4.4, there is a function $v(x, t)$ that is a periodic function of t, or a function $v(x)$ that does not depend explicitly on t, which satisfies all the conditions of the theorem to be proved with one possible exception. Suppose for definiteness that the function v has a positive-definite derivative dv/dt in the region H_0; then the doubtful property is that $v(x, t)$ take positive values in every neighborhood of the point $x = 0$. We have only to show that if the null solution $x = 0$ is unstable in $\Gamma, \bar{H}_0 \subset \Gamma$, the function $v(x, t)$ of theorem 4.3 also enjoys this doubtful property.

Suppose *per contra* that there is some neighborhood U, $0 \in U \subset H_0$ of the origin in which the function $v(x, t)$ is nonpositive. Suppose further (case 1) that for all $x \in U$, $v(x, t) < 0$ ($x \neq 0$). Consider a certain region $0 < \delta_1 \leq ||x||_2 \leq \delta_2$, contained in U. Since $v(x, t)$ is continuous and negative in this region, it will achieve a negative maximum there for $0 \leq t \leq \vartheta$. Thus, the periodic function $v(x, t)$ must be negative-definite in some neighborhood of the point $x = 0$. This would mean that the function v satisfies all the conditions of Lyapunov's theorem on asymptotic stability in this neighborhood, contradicting the hypothesis that the point $x = 0$ is unstable. This leads us to the second case. Suppose there is a sequence of points $x_0^{(k)} \neq 0$, and values $t_0^{(k)}, 0 \leq t_0^{(k)} < \vartheta$ such that $\lim x_0^{(k)} = 0$ and $v(x_0^{(k)}, t_0^{(k)}) = 0$. (In case the functions

X_i do not depend explicitly on the time, we take ϑ as an arbitrary positive number.) Since the derivative dv/dt is positive-definite along a trajectory of the system (1.3), we see that

$$v\big(x(x_0^{(k)}, t_k, 2\vartheta), 2\vartheta\big) = \int_{t_k}^{2\vartheta} \frac{dv}{dt}\, dt \,,$$

and this last integral exceeds the uniform positive bound

$$\int_{\vartheta}^{2\vartheta} \frac{dv}{dt}\, dt \,.$$

This is a contradiction, since $x(x_0^{(k)}, t_0^{(k)}, 2\vartheta) \to 0$ as $x_0^{(k)} \to 0$, which makes it impossible for v to satisfy $v \leqq 0$ in the neighborhood U. This contradiction shows that there must indeed be a sequence of points in which $v > 0$ converging on $x = 0$. Theorem 6.2 is proved.

7. Theorems on the Existence of a Function $v(x, t)$ that Satisfies the Conditions of Lyapunov's Second Theorem on Instability, or Četaev's Theorem on Instability

In this section we show that if $x = 0$ is unstable, it is always possible to construct a function $v(x, t)$ that satisfies the conditions of Lyapunov's second theorem on instability [114, theorem III, p. 226]. We also show that (if $x = 0$ is unstable) there is a function v that satisfies Četaev's theorem on instability [36, p. 34]. Thus both these theorems always give in principle a method of establishing the instability of the null solution $x = 0$ (when this solution is in fact unstable). We refer first to the theorem of Četaev, and then to Lyapunov's result. We make the following definition [36].

DEFINITION 7.1. *Let the function $v(x, t)$ be defined in a certain region H_0. We call the function dv/dt positive-definite in the region $v > 0$ (for $x \in H_0$, $0 < t < \infty$) provided that for every positive number $\delta > 0$ there is a positive number $\varepsilon = \varepsilon(\delta) > 0$ such that $dv/dt > \varepsilon$ for all x such that $x \in H_0$, $v(x, t) > \delta$.*

The direct theorem of Četaev reads as follows.

THEOREM (Četaev). *Suppose that a function $v(x, t)$ can be defined for all positive $t > 0$, having the property that for values of the arguments x_i that are arbitrarily small in absolute value $(i = 1, \cdots, n)$, the value of the derivative dv/dt along a trajectory of the equations (1.3) is positive-definite in the region $v > 0$. Then the perturbed motion is unstable* [36, p. 34].

We note that Četaev does not require the conditions to be satisfied in an entire neighborhood of the point $x = 0$. It is sufficient that the conditions are fulfilled only in the region $v > 0$. Moreover, the theorem will be valid if the region $v > 0$ contains an open set of points V such that $v(x, t) > 0$ for $x \in V$, and $v(x, t) \to 0$ for $x \to \bar{V} \setminus V$.

In Četaev's theorem, the set of points x, t for which $v(x, t) > 0$ is also an open set in the $(n + 1)$-dimensional space $x \times t$. The region V can become smaller with time, the change being continuous in t. If the solution $x = 0$

is unstable in a fixed region Γ (see footnote, p. 36), the theorem gives sufficient conditions that the function $dv(x, t)/dt$ be positive-definite in a region $\bar{H}_0 \subset \Gamma$. Moreover, the theorem has no content unless it is always possible to define a function v with the above properties for every H_0, $\bar{H}_0 \subset \Gamma$.

A converse of Četaev's theorem was first given for the case in which the functions $X_i(x, t)$ of equations (1.3) do not depend explicitly on the time [84]. Subsequently, Vrkoč [183] and the author [91] established the existence of a function $v(x, t)$ in the general case. Here we base our proof of the existence of a function v on lemma 3.1. The theorem we formulate is somewhat stronger than those in the references. We begin with a definition.

DEFINITION 7.2. *Let the null solution $x = 0$ of equations (1.3) be unstable in the region Γ, and let H_0 be a bounded region, $\bar{H}_0 \subset \Gamma$. By the region of instability for the null solution $x = 0$ in the region H_0 for $t = t_0$, we mean the set of points $E(t_0)$ such that the trajectory $x(x_0, t_0, t)$ always leaves the region $\bar{H}_0$ with increasing t whenever $x_0 \in E(t_0) \subset H_0$.*

Let H_0 be a given region, and suppose the region $\bar{H}_0$ satisfies the conditions of Četaev's theorem.* Then the points x_0 in the region $v > 0$ are contained in $E(t_0)$. We repeat Četaev's proof [36, pp. 34–35]. Suppose that $v(x_0, t_0) = \alpha > 0$. The relation $dv/dt > 0$ shows that the trajectory $x(x_0, t_0, t)$ remains in the region $v(x, t) > \alpha$ for all time, and hence in the region $v(x, t) > 0$. On the other hand $x(x_0, t_0, t) \in \bar{H}_0$ for $t \geq t_0$. Since the function dv/dt is positive in the region $v > 0$, $x(x_0, t_0, t)$ remains in the region $\bar{H}_0$. Thus the relation $dv/dt > \beta > 0$ holds for some positive constant β. Thus the estimate $v(x(x_0, t_0, t), t) > \beta(t - t_0)$ is valid for all such values of $t \geq t_0$. This shows that the trajectory $x(x_0, t_0, t)$ cannot lie in the region $\bar{H}_0$ for all values of t on the segment $t_0 \leq t \leq t_0 + T$, where $T = (\max |v| \text{ in } \bar{H}_0)/\beta$. The assertion is thus proved.

The question of a converse to Četaev's theorem comes down to the existence of a function $v(x, t)$ such that the region $v > 0$ is contained in the region of instability $E(t)$ of definition 7.2. We can give a satisfactorily complete answer to this question and show that a function $v(x, t)$ always exists with a region of instability that coincides with the region $E(t)$ in H_0. This is the assertion of the following theorem.

THEOREM 7.1. *Let the null solution $x = 0$ be unstable in the region Γ and let H_0 be a bounded region, $\bar{H}_0 \in \Gamma$. Then a function $v(x, t)$ can be defined that satisfies the conditions of Četaev's theorem on instability** in the region $\bar{H}_0$. That is, if x is a point in the region $v > 0$, $x \in \bar{H}_0$, the function dv/dt is positive-definite in the sense of definition 7.1, p. 39; the function v is bounded and*

---

* The derivative dv/dt along a trajectory of (1.3) is calculated from the formula $dv/dt = \sum X_i \partial v/\partial x_i + \partial v/\partial t$. Thus the partial derivatives in question must exist and be continuous in some neighborhood of the point at which dv/dt is being computed. The discussion requires that if a point $x \in \bar{H} \setminus H_0$ is such that $v(x, t) > \varepsilon > 0$, then for every $t > 0$ there be a neighborhood of points $x \in \bar{H}_0 \setminus H_0$ with the same property.

** See preceding footnote.

continuous in the region $\bar{H}_0$, and the partial derivatives $\partial v/\partial t, \partial v/\partial x_i$ $(i=1,\cdots,n)$ are bounded uniformly in the time. For every value of the time $t = t_0$, the region of instability $E(t_0)$ (in H_0) coincides with the region $v > 0$.

PROOF. We show first that the region of instability $E(t_0)$ is an open subset of $\bar{H}_0$. Take a point $x_0 \in E(t_0)$. The definition of $E(t_0)$ is such that there is a value $t = t^*$ for which the point $x(x_0, t_0, t^*) \in (\Gamma \backslash \bar{H}_0)$. The solution $x(x_0, t_0, t)$ depends continuously upon the initial values x_0, t_0. Thus we can find two positive numbers $\delta > 0, \tau > 0$ such that the points $x(x_0^*, t_0^*, t^*)$ all lie outside the region $\bar{H}_0$ whenever the inequalities $\| x_0^* - x_0 \| < \delta, \| t_0^* - t_0 \| < \tau$ are satisfied. The points $x_0^*, \| x_0^* - x_0 \| < \delta$ thus belong to the set $E(t_0)$. This is indeed the meaning of the assertion that the set $E(t_0)$ is open in $\bar{H}_0$. We have established more; we have shown that the set $[x_0, t_0]$ is open in the $(n + 1)$-dimensional space $x \times t$, where this set is defined as $E(t_0)$ for $0 < t_0 < \infty$. The collection of all such points x_0, t_0, where $x_0 \in E(t_0)$, $0 \leq t_0 < \infty$, we denote simply by E.

We now turn to the construction of the function v. We take $\delta_1 < 1$ and a monotone decreasing sequence of positive numbers $\delta_k, k = 1, 2, \cdots$. We denote by $E_k, k = 1, 2, \cdots$, the set of points $x \in \bar{H}_0, 0 \leq t < \infty$ defined by the relation

$$(7.1) \qquad \delta_k \leq t \leq k, \quad \rho[x, \{\bar{E}(t)\backslash E(t)\}] \geq \delta_k, \quad x \in E(t).$$

Let m be the first number for which the set E_m is nonempty. We consider the pairs x_0, t_0 contained in the regions E_k for $k \geq m$. The definition of E_k shows that the point x_0 is in $E(t)$. Thus we can find a point $t^* > t_0$ such that the point $x(x_0, t_0, t^*) \in \bar{H}_0$. Set $\gamma = \min \rho[x(x_0, t_0, t), \{\bar{H}_0\backslash E(t)\}]$ for $t_0 \leq t \leq t^*$.

We construct a function $v(x_0, t_0, x, t)$ that satisfies the conditions of lemma 3.1. This lemma shows that a function $v(x_0, t_0, x, t)$ does exist that satisfies the conditions

$$(7.2) \qquad v(x_0, t_0, x, t) = 0 \quad \text{in} \quad \bar{H}_0\backslash \bar{E}(t);$$

$$v > c, \, dv/dt > c$$

for some constant $c > 0$, when the relations $\| x_0 - x \| < \alpha, \| t_0 - t \| < \tau$ are satisfied, where α, τ are sufficiently small positive numbers; and the relation $dv(x_0, t_0, x(t), t)/dt \geq 0$ holds in the region $\bar{H}_0$.

The function $v(x_0, t_0, x, t)$ has continuous partial derivatives or arbitrary order with respect to all its arguments in the region Γ. We construct a function $v(x_0, t_0, x, t)$ of this type for every pair $(x_0, t_0) \in E_k$, where (x_0, t_0) runs through a finite number of pairs such that for that set of pairs the corresponding neighborhoods (7.2) cover E_k. Since E_k is a bounded closed set in $(n + 1)$-dimensional space $x \times t$, this is possible. We number this finite system of points with the indices $1, \cdots, N_k$ and denote the corresponding functions by the symbols $v_l^{(k)}(x, t), l = 1, \cdots, N_k$. We form the function

$$(7.3) \qquad v_k(x, t) = \sum_{l=1}^{N_k} v_l^{(k)}(x, t).$$

This function $v_k(x, t)$, $k = m, m + 1, \cdots$ clearly enjoys the following properties.

The function v_k is defined in the region $x \in \Gamma, 0 < t < \infty$, and possesses continuous partial derivatives of every order with respect to all its arguments in that region.

In the region $H_\delta \backslash E$ the relation $v_k = 0$ holds. For every $k \geq m$, moreover, we can find a number T_k such that $v_k = 0$ for $t \geq T_k$. This is true because each of the finite number of terms in the right member of (7.3) takes nonzero values only on a finite segment of the t-axis. Thus the modulus of the function v_k is bounded by some constant $P_k : |v_k| < P_k$.

Each of the partial derivatives

$$\frac{\partial^\mu v_k}{\partial x_1^{\mu_1} \cdots \partial x_n^{\mu_n} \partial t^{\mu_{n+1}}}$$

is bounded in modulus by some constant $P_k^{(\mu)}$.

In the region E_k, we have $dv_k/dt \geq \alpha > 0$; in the region $\bar{H}_0$, we have $dv_k/dt \geq 0$.

We define the function w_k by the equality $w_k(x, t) = v_k(x, t) \exp(t - T_k)$. Next we note that the relation

(7.4) $$\frac{dw_k}{dt} = w_k + \exp(t - T_k)\frac{dv_k}{dt}$$

holds. Now, since $dv_k/dt \geq 0$ in the region H_0, the function w_k satisfies the inequality $dw_k/dt \geq w_k$ in the region $\bar{H}_0$, and has all the other properties that the function v_k was shown to have. Thus, inequalities of the type

$$\left| \frac{dw_k}{\partial x_1^{\mu_1} \cdots \partial x_n^{\mu_n} \partial t^{\mu_{n+1}}} \right| < R_k$$

are valid for $x \in H_0, 0 < t < \infty$. The function $v(x, t)$ is defined by the relation

(7.5) $$v(x, t) = \sum_{k=m}^{\infty} \frac{w_k(x, t)}{(R_k 2^k)},$$

where R_k equals max $R_l^{(\mu)}$, the max being taken over the values $l = 1, \cdots, k$, $\mu = 1, \cdots, k$. This function $v(x, t)$ has all the properties required by the theorem. The function v is defined in the region $x \in \bar{H}_0, 0 < t < \infty$. The function v is zero: $v = 0$ in the region $\bar{H}_0 \backslash E(t)$ for every $t \geq 0$; also $v(x, t) > 0$, $dv/dt > v$ in the region $E(t)$. Thus the region $E(t)$ coincides with the region $v > 0$. The theorem is proved.

We emphasize that the proof of theorem 7.1 establishes more than a mere converse of Četaev's theorem; in fact, we show that this converse is equivalent to the converse of Lyapunov's second theorem on instability. This theorem was stated by Lyapunov [114, p. 266] in the following form.

THEOREM III. *Suppose a bounded function v can be defined so that the total derivative dv/dt of the function v along the trajectories of the system (1.3) has the form $dv/dt = \lambda v + w$, where λ is a positive number and w either is identically zero or is semidefinite. If w is not identically zero, suppose further that the function v has the property that for every t greater than a certain* *fixed bound the function v has the same sign as w for all sufficiently small values of x. Then the null solution is unstable.*

Remark. If we wish to conclude that the solution $x = 0$ is unstable in an entire region Γ, it is sufficient to require that a function v satisfying all conditions of theorem III should exist in every closed bounded subregion $\bar{H}_0 \in \Gamma$.

It is easy to see that the function v which was constructed in the proof of theorem 7.1 does satisfy all the conditions of theorem III of Lyapunov. We make this assertion in the form of the following theorem.

THEOREM 7.2. *If the null solution $x = 0$ of equation* (7.3) *is unstable in the region Γ, there is a function v that satisfies the conditions of Lyapunov's second theorem on instability in any preassigned closed bounded region $\bar{H}_0 \in \Gamma$. This function has continuous partial derivatives of all orders with respect to all arguments, and each derivative is bounded (with its own bound) in the region H_0, $0 < t < \infty$.*

We now consider the case in which the right members of equation (1.3), the functions X_i, are periodic functions of the time of period ϑ, or do not depend explicitly on the time. In this case it is interesting to show that there is a periodic function $v(x, t)$, or a function $v(x)$ that satisfies the conditions of Lyapunov's theorem on instability, or Četaev's theorem. The proof that such a function exists is obtained from the proof of theorem 7.2 by making modifications similar to those which came up in discussing the converse of the first theorem on instability (p. 36).

Suppose, for example, that the functions X_i are periodic functions of the time t of period ϑ. The function v is constructed as follows. The functions $w_k(x, t)$ are the functions $w_{kl}(x, t) = w_k(x, t + l\theta)$ $(l = 0, \pm 1, \pm 2, \cdots)$. The function w_k is

$$(7.6) \qquad W_k(x, t) = \sum_{l=-\infty}^{\infty} w_{kl}(x, t) .$$

Every term in formula (7.6) is nonzero for only a finite segment of values of t, and for each of the functions w_{kl} the length of this segment is uniformly bounded by the number l. Thus, in the neighborhood of every point $x \in \bar{H}_0$, $0 < t < \infty$, only a finite number of terms w_{kl} are different from zero, and this number is even uniformly bounded by a constant that is independent of the choice of x. Thus the functions W_k enjoy all the properties that the functions w_k did, including the estimates on the corresponding derivatives. Moreover, W_k is a periodic function of the time t with period ϑ.

The remaining steps in constructing the function v use the function W_k in the same way that the functions w_k were used earlier.

If the functions X_i are independent of the time t, we construct the function v by building first a sequence of functions $v_{l,k}(x, t)$, $l = 1, 2, \cdots$ of period ϑ/l, the details being analogous to the work above.

2

Some Modifications of Lyapunov's Theorems

8. Preliminary Remarks

In this chapter we shall describe various modifications of the classical theorems of Lyapunov. The idea of these modifications can be explained as follows: In Chapter 1 we showed that Lyapunov's theorems have a converse if the conditions of these theorems (including theorem III) involve not only the property of stability (asymptotic stability or instability) but also the property of *uniform* asymptotic stability ("property A"; see Chapter 1, p. 19). The connection between additional properties of the null solution of the differential system on the one hand, and additional restrictions on the Lyapunov function on the other hand, is a fundamental question. This is the essential question of a converse to the theorems. Modifications and generalizations of Lyapunov's theorems have been obtained by Četaev [36], K. P. Persidskii [145], [147], [148], Zubov [199], Massera [129]–[131], Kurcveil' [96], [97], [100], Gorbunov [61], and many others. We shall consider chiefly those generalizations and modifications which concern stability defined in a certain sense we call uniform. In a later section we give a modification of Četaev's theorem on instability, and show the connection between Četaev's theorem on instability and that of Persidskii [145], [147], [148], who gave criteria based on the concept *section*, which he introduced.

Those modifications and generalizations of Lyapunov's theorems which themselves admit converses are of special interest (and importance). By a converse, it will be recalled, we mean a theorem on the existence of a Lyapunov function, given a particular property of the trajectories. Only such theorems are discussed in this chapter.

9. Uniform Nonasymptotic Stability

The stability of the null solution $x = 0$ in the sense of definition 1.1 (p. 2) is given by Lyapunov's theorem [114, p. 82]:

THEOREM. *Suppose equations* (1.3) *are such that a positive-definite function* v *exists, and the total derivative* dv/dt *is negative-semidefinite along a trajectory of* (1.3). *Then the null solution is stable.*

This theorem is proved in the Introduction (p. 9). Persidskii [145], [147],

[148] showed that this theorem admits a converse. That is, if the functions X_i are supposed to have a certain degree of smoothness, there is always a function v that satisfies the conditions of the theorem, provided the null solution $x = 0$ is stable.

Persidskii's proof is as follows: Suppose the right members $X_i(x_1, \cdots, x_n, t)$ of equations (1.3) are defined and continuous in the region

(9.1) $$\| x \| < H, \quad 0 \leq t < \infty \quad (H = \text{const.}),$$

and suppose the partial derivatives $\partial X_i / \partial x_j$ $(i = 1, \cdots, n; j = 1, \cdots, n)$ are continuous in the same region. We suppose that the null solution $x = 0$ is stable in the sense of definition 1.1, p. 2, and prove the existence of a function $v(x_1, \cdots, x_n, t)$ satisfying the conditions of Lyapunov's theorem on stability in the region

(9.2) $$\| x \| < H_0, \ H_0 < H, \ 0 \leq t < \infty .$$

We define $Y_i(x, t) = X_i(x, t) \varphi(x)$, where φ is a continuous differentiable function that satisfies the conditions

(9.3) $$\varphi(x_1, \cdots, x_n) = 0 \quad \text{for} \quad \| x \| \geq H,$$

and

(9.4) $$\varphi(x_1, \cdots, x_n) = 1 \quad \text{for} \quad \| x \| < H_0 .$$

We consider the auxiliary system of equations

(9.5) $$\frac{dx_i}{dt} = Y_i(x_1, \cdots, x_n, t) \quad (i = 1, \cdots, n) .$$

We denote a solution of the system (9.5) by $y_i(x_0, t_0, t)$. The function v is defined by the relation

(9.6) $$v(x_1, \cdots, x_n, t) = (e^{-t} + 1) \sum_{i=1}^{n} y_i^2(x_1, \cdots, x_n, t, 0) .$$

We have to prove that this function satisfies the conditions of the theorem. The trajectory $y_i(x_0, t_0, t)$ has the following properties. When $\| x \| < H_0$, the system (9.5) coincides with the system (1.3). If we displace $x_i = x_i(t)$ $(i = 1, \cdots, n)$ along the trajectory $y_i(x_1, \cdots, x_n, t_0, t)$, (9.5) shows that the value of y_i in formula (9.6) does not change. (The term y_i in formula (9.6) is simply the initial value of the trajectory, i.e., the value at $t = 0$.) Thus $\partial v / \partial x_i = 0$ and we obtain the relation

$$\left(\frac{dv}{dt} \right)_{(1.3)} = -(\| y(x, t, 0) \|_2^2) e^{-t} < 0 \quad (x \neq 0) .$$

Thus the derivative dv/dt is a negative-definite function in the region (9.2). We claim further that the function (9.6) is positive-definite in the region (9.2). Indeed, let x_0 be a point in the region

(9.7) $$\varepsilon < \| x_0 \| < H_0 ,$$

and let $t > 0$ be some fixed moment of time. Since the solution is stable at $t_0 = 0$, we can find a number $\delta > 0$ such that the relation

$$|| x(x_0^*, 0, t) || < \varepsilon$$

holds whenever $|| x_0 || < \delta$. Thus the relation

(9.8) $$\delta < || y(x_0^*, t, 0) || < H$$

holds if the point x_0 is in the region (9.7). The second of these inequalities follows from the fact that the trajectory of the equation (9.5) which starts from the point $x_0 < H$ does not leave this region for $0 \leqq t < \infty$. This is true because the functions Y_i have been constructed so that the region is bounded by the singular points $|| x || = H$. Moreover, in the entire region $-\infty < x_i < \infty$, $0 \leqq t < \infty$ the system (9.5) has a unique solution. Inequality (9.8) shows that for the point x_0 being considered, the function (9.6) is indeed defined and satisfies the inequality $v(x_0, t) > \delta^2$. Thus the function v is positive-definite. The solution $y_i(x_{10}, \cdots, x_{n0}, t_0, t)$ is continuously differentiable with respect to the initial values x_{i0}. Therefore, the function (9.6) has continuous partial derivatives $\partial v / \partial x_i$ $(i = 1, \cdots, n)$. Thus the existence of a Lyapunov function v is established.

Remark. In general it is not possible to establish the existence of a function $v(x, t)$ which would have partial derivatives that are uniformly bounded in the region (9.2). If the functions X_i are periodic or do not depend explicitly on the time, it is not possible to show that a periodic function v exists, or that a function v exists which does not depend explicitly on the time. The difficulty is exemplified by a phase diagram (Fig. 3) for a system of two equations. In this figure the point $x = 0$ is surrounded by a countable system of trajectories, the concentric circles $Q_k, k = 1, 2, \cdots$. These circles approach the point $x = 0$ for $k \to \infty$.

The trajectories in the annular region bounded by the circles Q_{k+1}, Q_k are spirals that leave the circle Q_{k+1} and approach the circle Q_k. Suppose it were possible to construct a continuous function $v(x_1, x_2)$ defined in the neighborhood of the point $x_1 = x_2 = 0$, whose derivative dv/dt is negative-semidefinite along every trajectory. In view of the relation $dv/dt \leq 0$, the value of the function v at the point $x_1 = x_2 = 0$ cannot be less than the value of v on any one of the circles Q_k. The further requirement $v(0) = 0$ makes it clear that the function $v(x)$ cannot be positive-definite. This shows that in the case under consideration there is no function $v(x)$ of the arguments x_1, x_2 alone which would satisfy the conditions of Lyapunov's theorem on instability.

Let us now compare the properties of the Lyapunov function v in the theorem on stability with the function V of the theorem on asymptotic stability. The function V must satisfy two requirements in addition to those imposed on the function v:

(V1) The function V admits an infinitely small upper bound.

(V2) The derivative dV/dt is definite, whereas the derivative dv/dt is presumed to be only semidefinite.

Here we are assuming that the functions v and V are defined and satisfy

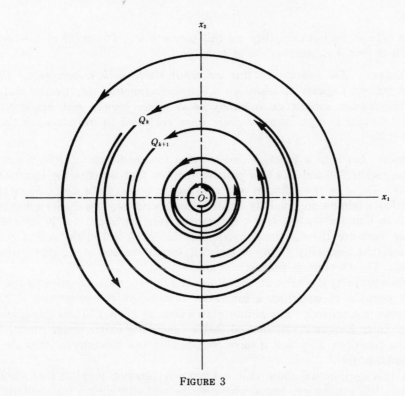

FIGURE 3

the conditions of the respective theorems in a region

(9.9) $$\| x \| < H, \ 0 \leqq t < \infty .$$

The additional properties V1 and V2 imply certain properties of the trajectories $x(x_0, t_0, t)$ of system (1.3). (See section 5, p. 26.)

(T1) The number $\delta > 0$ in definition 1.1, p. 2, is a function of the number ε, and independent of the initial time t_0.

(T2) The solution $x = 0$ is asymptotically stable.

(T3) In some region $\| x \| < H_0$, the trajectories $x(x_0, t_0, t)$ approach the solution $x = 0$ in a manner that is uniform with respect to the initial moment of time, $0 < t_0 < \infty$, and the initial value x_0.

It is also clear that property T1 is a consequence of property V1. This was first noted by K. P. Persidskii [147]. His definition can be formulated as follows.

DEFINITION 9.1. *The null solution $x = 0$ of equation (1.3) is called uniformly stable with respect to the initial time t_0, provided that there is a positive function $\delta = \delta(\varepsilon)$ of the positive argument $\varepsilon > 0$ such that for all t, t_0 with $t \geqq t_0 > 0$, the relation*

(9.10) $$\| x(x_0, t_0, t) \| < \varepsilon$$

holds for every initial value x_0 such that

(9.11) $|| x_0 || < \delta$.

This type of uniform stability we call property C. The point of this definition is that δ is independent of t_0.

THEOREM. *The solution $x = 0$ is uniformly stable with respect to t_0; that is, it enjoys property C whenever a Lyapunov function $v(x, t)$ exists that is positive-definite, admits an infinitely small upper bound, and has a total derivative that is negative-semidefinite when evaluated on trajectories of* (1.3). [See (1.13).]

PROOF. Let v be a function that satisfies the conditions of this theorem in the region (9.9) and let ε be a positive number that satisfies the inequality $\varepsilon < H$. Let λ be the infimum of $v(x, t)$ for $|| x || = \varepsilon$, $0 < t < \infty$. Since the function v admits an infinitely small upper bound, we can choose a number $\delta > 0$ so that for $|| x || \leqq \delta$, $0 < t < \infty$, the relation $v(x, t) < \lambda$ is satisfied. If we now use the additional hypothesis that $dv/dt \leqq 0$ in the region (9.9), we see that inequality (9.10) is satisfied for all values of x_0 that satisfy (9.11). The theorem is proved.

When property C is present, the interesting inverse question presents itself: is it possible to construct a function v that enjoys the properties of the theorem just proved? The author gave a positive answer to this question in [87]. Later Kurcveĭl' [96] showed that a function v exists under conditions on the functions X_i which involve considerably less smoothness than those assumed in [87].[*]

In this section we show that a Lyapunov function v exists that admits an infinitely small upper bound whenever the null solution $x = 0$ is uniformly stable in the sense of property C. We shall suppose that the functions X_i are defined, bounded, and continuous in a region (9.9). We shall suppose further that the derivatives $\partial X_i/\partial t$, $\partial X_i/\partial x_j$ ($i = 1, \cdots, n; j = 1, \cdots, n$) are continuous in this region. (The hypothesis that the derivative $\partial X_i/\partial t$ exist and be continuous can be omitted at the expense of complicating the proof.)

THEOREM 9.1. *If the null solution $x = 0$ of equation* (1.3) *is uniformly stable with respect to the time t_0, in the sense of property C, a function $v(x, t)$ can be defined that enjoys the following properties in a region* (9.2):
  (i) $v(x, t)$ *is positive-definite.*
  (ii) $v(x, t)$ *admits an infinitely small upper bound.*
  (iii) *The derivative dv/dt evaluated along the trajectories of* (1.3) *is negative-semidefinite.*
  (iv) $v(x, t)$ *has partial derivatives $\partial v/\partial t$, $\partial v/\partial x_i$ ($i = 1, \cdots, n$) that are continuous in the region* (9.2).

PROOF. The functions $X_i(x, t)$ can be extended into the region $-\theta < t \leqq 0$, for some positive constant $\theta > 0$, so that the extended functions are con-

---

[*] Further work on the existence of a Lyapunov function in the case of nonasymptotic stability is given in the combined paper of Kurcveĭl' and Vrkoč [100]. There it is not even assumed that equations (1.3) have a unique solution.

tinuously differentiable in the larger region. To do this, define $X(x, \tau)_{\tau < 0} = 2X(x, 0) - X(x, -\tau)$. Thus we assume from the beginning that the right members of equations (1.3) are defined and continuous and have continuous partial derivatives $\partial X_i / \partial t, \partial X_i / \partial x_j$ in a region

$$(9.12) \qquad ||\, x\, || < H, \ -\theta < t < \infty \ (\theta > 0)\,.$$

First we construct an auxiliary function $w(x, t)$ by means of the formula

$$(9.13) \qquad w(x_0, t) = (1 + e^{-t}) \inf [||\, x(x_0, t, \tau)\, ||_2]\,,$$

the infimum being taken over $0 \leqq \tau \leqq t$, where $x(x_0, t, \tau)$ is a point of a trajectory of the system (1.3),[*] $x(x_0, t, t) = x_0$. We assert that the function $w(x, t)$ is continuous in the region $||\, x\, || < H, \ 0 \leqq t < \infty$. The relation

$$\lim_{x_0' \to x_0, t_0' \to t_0} \ \big|\inf ||\, x(x_0, t_0, \tau)\, ||_2 - \inf ||\, x(x_0', t_0', \tau)\, ||_2\big| = 0$$

$$\text{for} \ \ 0 \leqq \tau \leqq t_0,\ 0 \leqq \tau \leqq t_0'$$

follows from the fact that the solution $x(x_0, t_0, t)$ depends continuously on the inital values. But this is the meaning of the assertion that the function $w(x, t)$ is continuous. Further, w is positive-definite and admits an infinitely small upper bound.

It follows from property C that if $||\, x_0\, || = \varepsilon > 0$, the relation

$$\inf ||\, x(x_0, t_0, t)\, ||_2 > \delta\,, \qquad 0 \leqq t \leqq t_0\,,$$

holds. Thus the inequality $w(x_0, t_0) > \delta / n$ is valid; that is, the function w is positive-definite. Moreover, it is clear that $\inf (||\, x(x_0, t_0, t)\, ||_2$ for $0 \leqq t \leqq t_0)$ does not exceed εn. Thus $w(x_0, t_0) \leqq \varepsilon 2n$, so that the function w admits an infinitely small upper bound. Next, the function $w(x, t)$ decreases steadily along a trajectory $x = x(t)$ as the time increases. This is obvious, since the second factor in the right member of formula (9.13) cannot increase as t increases, and the first factor decreases monotonically as t increases. Thus the function w satisfies all the conditions of the theorem to be proved except that it may not be sufficiently smooth and the value of the function $w(x(t), t)$ may not have a derivative dw/dt at isolated points. Thus, to conclude the proof of the theorem, it is necessary to construct a sufficiently smooth function $v(x, t)$ from the function w.

In fact, we show that the function w can be approximated by a smooth function v that possesses all the necessary properties.

Let H_1 be an upper bound of those numbers $\delta > 0$ for which the relation

$$||\, x(x_0, t_0, t)||_2 < H$$

holds whenever $||\, x_0\, ||_2 < \delta, \ 0 \leqq t_0 < \infty$. Property C shows that $H_1 > 0$. Let G_0 be some positive number such that $G_0 < H_1$. It is clear that the level surface $w(x, t) = G_0$ lies in the region $||\, x\, || < H$, since the relation

---

[*] Here we are considering system (9.5) instead of equations (1.3). To simplify the notation, we suppose that equations (1.3) have the properties needed, i.e., that the points $||\, x\, || = H$ are singular points of (1.3).

$$\limsup_{||x_0||\to H} \{\inf_{0\le t\le t_0} || x(x_0, t_0, t) ||_2\} \ge H_1$$

follows because of the way $w(x, t)$ was constructed and the way the number H_1 was chosen.

Let δ_k $(k = 0, 1, 2, \cdots)$ be a monotonically decreasing sequence of numbers, $\delta_0 = G_0$, $\lim \delta_k = 0$. It is clear that the trace of the level surface $W = \delta_k$ by the plane $t = 0$ is the sphere $|| x ||_2^2 = \delta_k^2$. Let us suppose the number k to be fixed and construct the region h_k in the $(n + 1)$-dimensional $x \times t$ space in the following manner.

For $t > 0$ the point (x, t) is contained in the set H_k if $\delta_{k+1} < w(x, t) < \delta_k$; for $t < 0$ the point (x, t) is contained in the set h_k if one of the following conditions is satisfied:

$$\text{for} \quad \frac{-\theta}{k + 1} < t < 0, \qquad \delta_{k+1} < || x(x, t, 0) ||_2 < \delta_k ,$$

$$\text{or for} \quad \frac{-\theta}{k + 1} < t < \frac{-\theta}{k + 2} \quad || x(x, t, 0) ||_2 < \delta_k .$$

The regions h_k $(k = 0, 1, 2, \cdots)$ constructed in this manner do not intersect. We show that each such region h_k is a *section* [138, p. 500]. That is, in the region h_k there is a surface S_k that is the level surface of some continuously differentiable function $u_k(x, t)$, and every arc $x(x_0, t_0, t)$ of the region h_k pierces S_k once and only once. (By an arc $x(x_0, t_0, t)$ of the region h_k we mean a maximal connected arc of the trajectory (1.3) that lies wholly in the region h_k.) The surface S_k is a *smooth section* [14] if the inequality

$$(9.14) \qquad \sum_{i=1}^{n} \frac{\partial u_k}{\partial x_i} X_i + \frac{\partial u_k}{\partial t} > 0$$

is satisfied.

Barbašin showed [14] that such a smooth section S_k always exists if the system of arcs that fills out the region h_k is *totally unstable*. The region h_k is said to be totally unstable, if whenever a closed region r is contained in h_k, every arc on h_k will forsake this closed region with both increasing and decreasing time. It is clear that the system of arcs that fills out the surface h_k we have constructed is totally unstable; thus a smooth section S_k must exist.

The sections S_k are the means for constructing a smooth function $v(x, t)$ that satisfies the conditions of the theorem.

For every integer $k \ge 0$ we can find a number $\tau_k > 0$ with the property that if $(x_0, t_0) \in S_k$, a segment of an arbitrary trajectory $x(x_0, t_0, t)$ will not intersect the point $x = 0$ if $t \ge 0$ and $t_0 \le t \le t_0 + \tau_k$. To demonstrate this let min $\rho[O, S_k] = \nu_k$. This quantity is positive, $\nu_k > 0$, because of the way the section $S_k \subset h_k$ was constructed. (Here ρ is defined in the product space $x \times t$, and O is the axis $x = 0, t \ge 0$.) Thus the speed with which the point $x(x_0, t_0, t)$ moves along the trajectory in the space x in the region

$||x|| < H$ is uniformly bounded, since the functions X_i are bounded in the region (9.9). We now define the continuously differentiable function $\varphi_k(\tau)$ by the conditions

(9.15)
$$\varphi_k(\tau) = 1 \quad \text{for} \quad \tau \leq \frac{\tau_k}{2},$$
$$\varphi_k(\tau) = 0 \quad \text{for} \quad \tau \geq \tau_k,$$
$$\varphi_k'(\tau) < 0 \quad \text{for} \quad \frac{\tau_k}{2} < \tau < \tau_k.$$

The functions φ_k and the sections S_k can be used to construct the terms $v_k(x, t)$ and from them the function v. Let k be fixed and suppose that $||x_0|| < H$ for $t_0 \geq 0$. We consider the following two mutually exclusive cases.

(1) The trajectory $x(x_0, t_0, t)$ pierces the section S_k for $t = t_s \leq t_0$. In this case we define

(9.16)
$$v_k(x_0, t_0) = \varphi_k(t_0 - t_s).$$

(2) The trajectory $x(x_0, t_0, t)$ does not intersect the section S_k for $t \leq t_0$. In this case we define

(9.17)
$$v_k(x_0, t_0) = 1.$$

The function v_k that we have defined enjoys the following properties: it is defined in every point $||x|| < H$ for $t > 0$; and, for every $k \geq 0$, we can find numbers $\Delta_k > 0$ and η_k such that the inequalities

(9.18)
$$v_k(x, t) = 1 \quad \text{for} \quad ||x|| \geq \Delta_k,$$
$$v_k(x, t) = 0 \quad \text{for} \quad ||x|| \leq \eta_k$$

hold for $t > 0$.

The functions $v_k(x, t)$ do not increase as the point x moves along the trajectory $x = x(x_0, t_0, t)$. This is true because while the point $x(t)$ moves along the trajectory before reaching the section S_k we have $v_k \equiv 1$, and after crossing the section S_k with increasing time we have

$$\frac{dv_k}{dt} = \frac{d\varphi_k(t - t_s)}{dt} \leq 0$$

by relation (9.15).

Now let us show that the $v_k(x, t)$ have continuous partial derivatives $\partial v / \partial x_i$, $\partial v_k / \partial t$ $(i = 1, \cdots, n)$. Let t_s be the moment at which the trajectory $x(x_0, t_0, t)$ pierces the section S_k. It is clearly sufficient to check that v_k is differentiable in the points $x = x_0$, $t = t_0$, at which $t_0 - t_s > 0$. To do this we show that $t_s(x_0, t_0)$ is continuous and differentiable as a function of the arguments $x_{10}, \cdots, x_{n0}, t_0$. Suppose that when x_i increases from x_{i0} to $x_{i0} + \Delta x_{i0}$, with $x_j = x_{j0}$ for $j \neq i$, t_s increases by Δt_s. If the pair (x_0, t_0) lies in the region where the trajectory $x(x_0, t_0, t)$ pierces the section S_k for $t < t_0$, then it will automatically be true that the pair (x_0', t_0') lies in the same region if the quantities $||x_0' - x_0||$ and $||t_0' - t_0||$ are sufficiently small.

This is true because the trajectory $x(x_0, t_0, t)$ intersects S_k for $t < t_0$. Thus for some t less than t_0 there is a point on this trajectory that lies in the region h_k, where a function u_k is defined, whose level surface $u_k = 0$ is the section S_k. There are two points t_1 and t_2 lying on the trajectory $x(x_0, t_0, t)$ at which the function u_k has opposite signs: $u_k(x(x_0, t_0, t_1), t_1) < 0$, $u_k(x(x_0, t_0, t_2), t_2) > 0$. Since the functions $u_k(x, t)$ are continuous in x and t, and since the solutions of the differential equations are continuous functions of the initial conditions, we have $u_k(x(x_0', t_0', t_1), t_1) < 0$ and $u_k(x(x_0', t_0', t_2), t_2) > 0$ for all values of x_0', t_0' in a sufficiently small neighborhood of the point x_0, t_0. This makes it clear that the trajectory $x(x_0', t_0', t)$ cuts the section S_k for some $t_1 < t < t_2$. Since the numbers t_1 and t_2 can be chosen arbitrarily close to one another, it is apparent that the function $t_S(x_0, t_0)$ is continuous. Let $x_i^{(S)} = x_i(x_0, t_0, t_S(x_0, t_0))$. The solutions are differentiable functions of the initial conditions, so that we can write, for the above values Δx_{i0},

$$(9.19) \qquad \Delta x_j^{(S)} = \frac{\partial \tilde{x}_j}{\partial x_{j0}} \Delta x_{i0} + \frac{\partial \tilde{x}_j}{\partial t_S} \Delta t_S \qquad (j = 1, \cdots, n),$$

where the symbol $\sim$ denotes that the derivative in question is calculated at an intermediate point. Furthermore, the theorem of the mean can be used as follows:

$$(9.20) \qquad u_k(x_1^{(S)} + \Delta x_1^{(S)}, \cdots, x_n^{(S)} + \Delta x_n^{(S)}, t_S + \Delta t_S) - u_k(x_1^{(S)}, \cdots, x_n^{(S)}, t_S)$$
$$= \sum_{j=1}^n \frac{\partial \tilde{u}_k}{\partial x_j} \Delta x_j^{(S)} + \frac{\partial \tilde{u}_k}{\partial t} \Delta t_S = 0,$$

where the derivatives $\partial u_k / \partial x_j$, $\partial u_k / \partial t$ are also evaluated at an intermediate point. From equations (9.19) and (9.20) we obtain the relation

$$(9.21) \qquad \frac{\Delta t_S}{\Delta x_{i0}} \left(\sum_{j=1}^n X_j \frac{\partial \tilde{u}_k}{\partial x_j} + \frac{\partial \tilde{u}_k}{\partial t} \right) = - \sum_{j=1}^n \frac{\partial \tilde{u}_k}{\partial x_j} \frac{\partial \tilde{x}_j}{\partial x_{i0}}.$$

By inequality (9.14) the second factor in the left member of equation (9.21) has a nonzero limit. Thus equation (9.21) shows that the function $t_S(x_0, t_0)$ is a differentiable function, and that moreover,

$$(9.22) \qquad \frac{\partial t_S}{\partial x_{i0}} = - \left(\sum_{j=1}^n \frac{\partial u_k}{\partial x_j} \frac{\partial x_j}{\partial x_{i0}} \right) \Big/ \left(\sum_{j=1}^n \frac{\partial u_k}{\partial x_j} + \frac{\partial u_k}{\partial t} \right).$$

Thus it is seen that the function t_S is continuously differentiable with respect to x_0. In the same way we see that t_S is differentiable with respect to t_0 as well. The details need not be given. Thus the functions $v_k(x, t)$ are continuously differentiable in the region (9.9).

We take a sequence of unbounded monotonically increasing numbers $T_k (k = 0, 1, 2, \cdots)$. We choose positive numbers N_k that satisfy the inequalities

$$(9.23) \qquad \left| \frac{\partial v_m}{\partial x_j} \right| < N_k, \qquad |v_m| < N_k, \qquad \left| \frac{\partial v_m}{\partial t} \right| < N_k$$

$$\text{for } m \leq k, \, 0 \leq t \leq T_k, |x| < H.$$

The function

$$(9.24) \qquad v(x_1, \cdots, x_n, t) = \sum_{k=0}^{\infty} (1 + e^{-t}) v_k(x_1, \cdots, x_n, t) \frac{1}{N_k 2^k}$$

satisfies all the conditions of theorem 9.1:

(i) The function v is positive-definite, since for $\| x \| = \varepsilon$, we have

$$v(x, t) \geq \sum_{k=K}^{\infty} \frac{1}{N_k 2^k} = \gamma > 0 \,,$$

where $K > 0$ is a positive number, namely the smallest number k that satisfies the inequality $\varepsilon > \varDelta_k$.

(ii) The function v admits an infinitely small upper bound, since for $\| x \| = \varepsilon$ the relation

$$v(x, t) = \sum_{k=K_1}^{\infty} \frac{2}{N_k 2^k} = \omega < \infty$$

holds. Here K_1 is the smallest (positive) number k for which the relation $\varepsilon < \eta_k$ is satisfied.

(iii) The function v has continuous partial derivatives $\partial v / \partial x_i$, $\partial v / \partial t$, since the series (9.24) and the series obtained by differentiating (9.24) converge absolutely and uniformly, by choice of the numbers N_k. As a function of the time, every term in the right member of (9.24) decreases monotonically when the point $x = x(t)$ moves in region (9.2) along a trajectory $x(x_0, t_0, t)$. Thus the function $v(x(x_0, t_0, t), t)$ has a negative-semidefinite derivative in that region. The theorem is proved.

It should be remarked that it is not in general true that the partial derivatives $\partial v / \partial x_i$, $\partial v / \partial t$ are uniformly bounded in the entire region (9.2).

10. Generalizations of Lyapunov's Theorems on Asymptotic Stability when Stability is Nonuniform

The theorem demonstrated in the preceding section is intermediate between Lyapunov's theorems on stability and those on asymptotic stability. This is clear when we recall that the additional property of uniform stability in the sense of definition 9.1 is added to the property of ordinary stability in that theorem. Asymptotic stability as such is not involved. Additional assumptions must be made concerning the function v in the last two theorems of the preceding section in order to assure that the stability will be asymptotic, that is, that the trajectories will approach the point $x = 0$. Lyapunov's theorem on asymptotic stability does impose the additional condition that the derivative dv/dt be negative-definite. This hypothesis guarantees not only that the trajectory $x(x_0, t_0, t)$ will approach the null solution $x = 0$, but even that this approach will be uniform with respect to t_0 and x_0. For nonuniform approach of the perturbed trajectories to the point $x = 0$, a less stringent requirement on the derivative dv/dt is permissible. In this section we establish such a theorem and a converse for it.

Let us recall the precise definition of uniform asymptotic stability. In section 5 (p. 25) asymptotic stability was defined, uniform with respect to t_0 and x_0 (property B). At the beginning of the book (definition 1.2, p. 3) a definition of asymptotic stability was given, not uniform with respect to the coordinate x_0 or the time t_0. But we can suppose, for example, that all the conditions of definition 1.2 may be satisfied and that stability is moreover uniform with respect to the coordinates x_0. The following definition makes this precise.

DEFINITION 10.1. *The null solution $x = 0$ of equation* (1.3) *is called asymptotically stable uniformly with respect to the coordinates $x_0 \in H_0$ if all the conditions of definition 1.2 are satisfied, and the following condition is also satisfied: For arbitrary $\eta > 0$ and $t_0 \geqq 0$, there is a number $T(\eta, t_0)$ such that if $t \geqq t_0 + T(\eta, t_0)$,*

$$\| x(x_0, t_0, t) \| < \eta$$

for all values of the initial point $x_0 \in H_0$. In this case we shall say that the trajectories have property D.

Properties C and D are not independent. In general, property D may occur with or without property C. If the stability is asymptotic, property D is a consequence of property C.

LEMMA 10.1. *Let the null solution $x = 0$ be asymptotically stable, and suppose the closed, bounded region $\bar{H}_0$ lies in the region of attraction of the null solution. If the stability is uniform with respect to t_0 in the sense of property C, then the asymptotic stability is uniform with respect to the coordinates $x_0 \in \bar{H}_0$, in the sense of definition 10.1, property D.*

PROOF. Let the number $\eta > 0$ be given. By hypothesis, stability is uniform in the sense of property C. Thus there is a number $\delta = \delta(\eta) > 0$ such that whenever $\| x_0 \| < \delta$, the relation

(10.1) $$\| x(x_0, t_1, t) \| < \eta$$

holds for all $t \geqq t_1 \geqq 0$. Since the stability is asymptotic whenever $x_0 \in \bar{H}_0$, it will be possible to choose a number $t(x_0)$ so large that

$$\| x(x_0, t_0, t(x_0)) \| = \frac{\delta}{2} .$$

Now the solution depends continuously on the initial conditions. Thus we can find another number $\gamma(x_0) > 0$ such that whenever

(10.2) $$\| x_0' - x_0 \| < \gamma(x_0) ,$$

the relation

$$\| x(x_0', t_0, t(x_0)) \| < \delta , \qquad x(x_0', t_0, t) \in H$$

holds for $t_0 \leqq t \leqq t(x_0)$. The bounded closed set $\bar{H}_0$ can be covered by a finite system of neighborhoods of the form (10.2). Therefore, there will be a number T $[T = \max t(x_0)]$ such that every trajectory $x(x_0, t_0, t)$ enters the region $\| x \| < \delta$ as t increases on the segment $t_0 \leq t \leq T$. This then proves the lemma, because inequality (10.1) shows that the number δ has been chosen so that $\| x(x_0, t_0, t) \| < \eta$ for all $t \geq T$.

We shall confine our considerations to asymptotic stability that is uniform in the coordinates in the sense of definition 10.1, property D. The following theorem gives a sufficient condition for asymptotic stability of this type.

THEOREM 10.1. *Suppose, for the differential equations of perturbed motion, that one can define a function $v(x, t)$ in the region $x \in H$ with the following properties:*

(i) *v admits an infinitely small upper bound in H;*

(ii) *v is positive-definite in H;*

(iii) *v satisfies the condition*

$$(10.3) \qquad \frac{dv}{dt} \leq 0 \ in \ the \ region \ H;$$

(iv)

$$(10.4) \qquad \sup [v(x, t) \ in \ the \ region \ H_0, \ 0 \leq t < \infty]$$
$$< [\inf v(x, t) \ on \ the \ boundary \ of \ H_1, 0 \leq t < \infty] \,,$$

where the regions H_0, H_1, H satisfy the inclusions $\bar{H}_0 \subset H_1$, $\bar{H}_1 \subset H$;

(v)

$$(10.5) \qquad \int^{\infty} m_\eta(\tau) \, d\tau = -\infty$$

for all sufficiently small positive $\eta > 0$, where $m_\eta(t)$ is the supremum o dv/dt in the region $\eta \leq \| x \|, x \in H_1$.

Then the null solution $x = 0$ is asymptotically stable uniformly with respect to the coordinates x_0 in the sense of definition 10.1, property D, and is also uniformly stable in the sense of definition 9.1, property C.

PROOF. The uniform stability of the null solution $x = 0$ follows from the theorem of section 9 (p. 48). Thus all that is required is to prove that the solution is asymptotically stable. Let $\eta > 0$ be an arbitrary positive number, and let $\delta = \delta(\eta)$ satisfy condition (10.1). Set $N = \max |v|$ in the region $x \in H_1$, $0 \leq t$. By condition (10.5), for every $t_0 = 0$, one can find a number T such that

$$\int_{t_0}^{t_0 + T} m_\delta(\tau) \, d\tau < -N \,.$$

Let $x_0 \in H_0$. Then the estimate

$$v(x(x_0, t_0, t), t) - v(x_0, t_0) \leq \int_{t_0}^{t} \left(\frac{dv}{dt} \right) dt \leq \int_{t_0}^{t} m_\delta(\tau) \, d\tau$$

is valid for all $t > t_0$ for which the trajectories remain constantly outside the region $\|x\| < \delta$. However, if the inequality $\|x(x_0, t_0, t)\| > \delta$ were satisfied on the segment $t_0 \leq t \leq t_0 + T$, the relation

$$v(x(x_0, t_0, t), t) \leq N + \int_{t_0}^{t_0+T} m_8(\tau)\, d\tau < 0$$

would be valid, and this is not possible, since v is a positive-definite function. This contradiction shows that every trajectory $x(x_0, t_0, t)$ does contain a point in the region $\|x\| < \delta$ for some value of the time t on the segment $t_0 \leq t \leq t_0 + T$. Now by the choice of the number δ, it follows that the inequality $\|x(x_0, t_0, t)\| < \eta$ is satisfied for all $t > t_0 + T$. Thus the assertion of the theorem is established.

The theorem just proved admits a valid converse similar to the converse of Lyapunov's theorem on asymptotic stability.

THEOREM 10.2. *Let the null solution $x = 0$ be asymptotically stable uniformly with respect to the coordinate $\|x_0\| < H_1$ in the sense of property D, definition 10.1, where H_1 is some positive constant. Suppose further that the stability is uniform in the sense of property C, definition 9.1. Then one can define a function $v(x, t)$ in the region $\|x\| < H_1$, $0 \leq t < \infty$ that has the properties mentioned in theorem 10.1:*

(i) *v is positive-definite.*
(ii) *v admits an infinitely small upper bound.*
(iii) *The derivative dv/dt is negative-semidefinite in the region $\|x\| < H_1$.*
(iv) *The function dv/dt satisfies condition (10.5).*
(v) *The function $v(x, t)$ has continuous partial derivatives $\partial v/\partial t$, $\partial v/\partial x_i$ $(i = 1, \cdots, n)$, $0 \leq t < \infty$, $\|x\| < H_1$.*

The proof of theorem 10.2 is not included here, since it follows the line of proof of similar theorems. It should be mentioned that theorems 10.1 and 10.2 admit a certain modification that includes asymptotic stability when the stability is not uniform in the sense of definition 10.1, property D. The formulation of such a theorem was given by the author [93].

In concluding this section, we call attention to the fact that if $p_{ij}(t)$ are continuous and bounded functions of the time t, then the linear equations

(10.6)
$$\frac{dx_i}{dt} = p_{i1}(t)x_1 + \cdots + p_{in}(t)x_n$$

of perturbed motion are always stable uniformly in the initial coordinates x_0 in the sense of definition 10.1, property D. In this case the theorem just proved and its converse take the following form.

THEOREM 10.3. *Necessary and sufficient conditions that the solutions of system (10.6) be asymptotically stable are that a function $v(x, t)$ can be defined with the following properties.*

(i)

(10.7)
$$v(x, t) = \sum_{i,j=1}^{n} a_{ij}(t)\, x_i x_j,$$

(ii) $v(x, t)$ *is positive-definite.*

(iii) *The derivative*

(10.8)
$$\frac{dv}{dt} = \sum_{i,j=1}^{n} c_{ij}(t)\, x_i x_j$$

satisfies the relation

(10.9)
$$\frac{dv}{dt} \leq \lambda(t)\, v ,$$

where $c_{ij}(t) = \sum_{k=1}^{n} a_{ik} p_{kj} + a_{jk} p_{ki} + a_{ij}'(t)$, *and*

(10.10)
$$\int^{\infty} \lambda(t)\, dt = -\infty .$$

Remark. In formulating this theorem the requirement that the derivative dv/dt be negative-semidefinite is not explicitly made. That is, the value of $\lambda(t)$ in inequality (10.9) may be positive at times. Since the derivative dv/dt and the function $v(x, t)$ are quadratic forms in the case under consideration, it follows in particular that the number $\lambda(t)$ may be taken to be the largest value of the function dv/dt (10.8) on the surface

(10.11)
$$v(x, t) = \sum_{i,j=1}^{n} a_{ij}(t) x_i x_j = 1 .$$

The maximum value of dv/dt for $v = 1$ will be the greatest root of the determinantal equation

(10.12)
$$\det \begin{bmatrix} c_{11} - \lambda a_{11} \cdots c_{1n} - \lambda a_{1n} \\ \cdots \cdots \cdots \cdots \\ c_{n1} - \lambda a_{n1} \cdots c_{nn} - \lambda a_{nn} \end{bmatrix} = 0$$

according to the classical theory of quadratic forms. Applying a result of Razumihin [158], it follows that the solution is asymptotically stable. Razumihin's criterion applies to linear systems and amounts to the requirement that the greatest root of equation (10.12) satisfy condition (10.10). Another proof of this assertion appears in the work of Gorbunov [61]. In any case, the result follows at once from the relation

$$v(x, t) \leq v(x_0, t_0) \exp \int \lambda(t)\, dt ,$$

which is obtained by integrating (10.9).

11. Lyapunov Functions that Satisfy Estimates Given in Terms of Quadratic Forms

There are problems in which it is important to be able to construct a Lyapunov function v that not only enjoys the properties thus far mentioned (positive-definiteness, admitting an infinitely small upper bound), but also enjoys some further properties. Among the most important of such Lyapunov functions are those which satisfy estimates characterized by quadratic forms.

We begin with known facts about the system of differential equations with constant coefficients

(11.1) $\dfrac{dx_i}{dt} = d_{i1}x_1 + \cdots + d_{in}x_n \qquad (i = 1, \cdots, n)$.

This system has the trivial solution $x = 0$. The trivial solution is asymptotically stable if and only if the roots λ_i $(i = 1, \cdots, n)$ of the characteristic equation

(11.2) $\det \begin{bmatrix} d_{11} - \lambda \cdots d_{1n} \\ \cdot \ \cdot \ \cdot \ \cdot \ \cdot \ \cdot \ \cdot \\ d_{n1} \cdots d_{nn} - \lambda \end{bmatrix} = 0$

of the system all have negative real part. This means that there is a positive constant δ such that

(11.3) $\operatorname{Re} \lambda_i < -\delta \qquad (\delta = \text{const.}; \ \delta > 0; \ i = 1, \cdots, n)$.

See [114, p. 252].

In the same work [114, p. 276], Lyapunov showed that whenever condition (11.3) is satisfied, if $\sum c_{ij}x_ix_j$ is any preassigned quadratic form, there will be a quadratic form

(11.4) $v(x_1, \cdots, x_n) = \sum\limits_{i,j=1}^{n} a_{ij}x_ix_j \qquad (a_{ij} = a_{ji})$

such that the condition

(11.5) $\dfrac{dv}{dt} = \sum\limits_{i=1}^{n} \dfrac{\partial v}{\partial x_i}(d_{i1}x_1 + \cdots + d_{in}x_n)$

$= \sum\limits_{i,j=1}^{n} c_{ij}x_ix_j$

$= w(x_1, \cdots, x_n) \qquad (c_{ij} = c_{ji})$

is satisfied. See also [144].

If the quadratic form in the right member of equation (11.5) is negative-definite, the quadratic form v must be positive-definite.

It is also to be noted that the coefficients c_{ij} in equation (11.5) are given by the formula

(11.6) $c_{ij} = \sum\limits_{k=1}^{n} (a_{ik}d_{kj} + a_{jk}d_{ki})$.

Thus the matrix $C = (c_{ij})$ that defines the quadratic form dv/dt is the symmetric matrix product of the matrices $A = (a_{ij})$ and $D = (d_{ij})$.

We shall use the nomenclature *negative-definite* for the quadratic form $w(x_1, \cdots, x_n)$ of (11.5). Thus we can find positive constants $c_1, \cdots, c_4$ such that

$c_1 \| x \|_2^2 \leqq v(x) \leqq c_2 \| x \|_2^2$,

(11.7) $\dfrac{dv}{dt} = w(x) \leqq -c_3 \| x \|_2^2$,

$\left| \dfrac{\partial v}{\partial x_i} \right| \leqq c_4 \| x \|_2$.

Thus if the linear system (11.1) is asymptotically stable, a Lyapunov function always exists, namely, a quadratic form v that satisfies the conditions (11.7).

The above result can be carried over to linear systems with variable coefficients

$$(11.8) \qquad \frac{dx_i}{dt} = p_{i1}(t)x_1 + \cdots + p_{in}(t)x_n \qquad (i = 1, \cdots, n),$$

where $p_{ij}(t)$ are continuous and bounded on $0 < t < \infty$. The generalizations in question are due to K. P. Persidskii [145] and Malkin [121]; we shall now summarize some of their results.

If there is a Lyapunov function v that satisfies conditions (11.7) (we do not require that the function v be a quadratic form or that the system of equations be linear), not only must the solution $x = 0$ be asymptotically stable, but also the inequality

$$(11.9) \qquad || x(x_0, t_0, t) ||_2 \leq B || x_0 ||_2 \exp [-\alpha(t - t_0)]$$

must hold for $t \geq t_0$. Here, α, B are positive constants.

This can be seen as follows. The relation $dv/dt \leq -c_3 v/c_2$ follows from (11.7). Integrating this, we obtain

$$v(x(x_0, t_0, t), t) \leq v(x_0, t_0) \exp \left[-\frac{c_3}{c_2}(t - t_0) \right].$$

By using estimate (11.7) again, we obtain

$$(11.10) \qquad || x(x_0, t_0, t) ||_2^2 \leq \frac{c_2}{c_1} || x_0 ||_2^2 \exp \left[-\frac{c_3}{c_2}(t - t_0) \right],$$

which establishes (11.9).

If the system of equations is linear with constant coefficients and stability is asymptotic, relation (11.9) is always satisfied. This is true because the function $v(x)$ exists and satisfies relations (11.7). For a system with variable coefficients, relations (11.7) are not generally satisfied in the case of asymptotic stability; still less is it to be expected in more general cases that there will be a Lyapunov function $v(x, t)$ that satisfies relations (11.7). The question naturally arises whether one can conclude that if relations (11.9) are satisfied there must be a function $v(x, t)$ that satisfies (11.7). For linear systems with variable coefficients this is indeed the case.

Indeed, Persidskii [145] and Malkin [121] proved that if the system (11.8) possesses a function $v(x, t)$ that satisfies the conditions of Lyapunov's theorem on asymptotic stability, then the null solution $x = 0$ of this system satisfies relation (11.9). Conversely, if a solution $x(x_0, t_0, t)$ of the system (11.8) satisfies condition (11.9), then there is a quadratic form $v(x, t)$ in which the coefficients $a_{ij}(t)$ are variable, and for which the estimates (11.7) are valid.

We see from the preceding discussion that conditions (11.9) follow from the existence of a function $v(x, t)$ that satisfies the estimate (11.7), whether the system is linear or not. Indeed the linearity of the system was not used in the proof. A more general result of this nature is proved in the following paragraph, where the function v itself is not required to be a quadratic form.

In this section we assume that the functions $X_i(x, t)$ in the right member of the equations

(11.11) $$\frac{dx_i}{dt} = X_i(x_1, \cdots, x_n, t) \qquad (i = 1, \cdots, n)$$

of perturbed motion are continuous, and have continuous partial derivatives $\partial X_i/\partial x_j$ $(i = 1, \cdots, n;\ j = 1, \cdots, n)$ in the region

(11.12) $$\| x \|_2 < H, \quad 0 < t < \infty,$$

where $H = \text{const.}$ or $H = \infty$. Moreover, we assume that the estimates

(11.13) $$\left| \frac{\partial X_i}{\partial x_j} \right| < L \quad (L = \text{const.};\ i, j = 1, \cdots, n)$$

are also valid in this region.

Let B, α, and H_0 be positive constants with $B > 0$, $\alpha > 0$, $H_0 \leq H/B$. We assume that when the initial conditions x_0 satisfy the relations

(11.14) $$\| x_0 \|_2 < H_0, \qquad t_0 \geq 0$$

the trajectories $x(x_0, t_0, t)$ of the system (11.11) satisfy the relation (11.9), that is,

(11.15) $$\| x(x_0, t_0, t) \|_2 \leq B \| x_0 \|_2 \exp[-\alpha(t - t_0)] \quad \text{for} \quad t \geq t_0.$$

THEOREM 11.1. *Whenever the solution $x(x_0, t_0, t)$ of the system* (11.11) *satisfies relation* (11.15), *then there is a function $v(x, t)$ that is defined in the region $\| x \| < H_0$ and satisfies the estimates* (11.7) *there.*

PROOF. Set $T = [\ln(B\sqrt{2})]/\alpha$. We define the function $v(x, t)$ by the formula

(11.16) $$v(x_0, t_0) = \int_{t_0}^{t_0+T} \| x(x_0, t_0, t) \|_2^2 \, dt,$$

and show that it satisfies all the conditions of the theorem.

(1) The function $v(x, t)$ is defined in the region (11.14). Indeed, relation (11.15) asserts that the solution $x(x_0, t_0, t)$ does not have the region (11.12) for $t \geq t_0$. From the same relation, the estimate

$$v(x_0, t_0) \leq \int_{t_0}^{t_0+T} \| x_0 \|_2^2 B^2 \exp[-2\alpha(t - t_0)] \, dt = c_2 \| x_0 \|_2^2$$

is obtained.

(2) From inequality (4.11) we obtain the estimate

$$\| x(x_0, t_0, t) \|_2 \geq \| x_0 \|_2 \exp[-nL(t - t_0)], \quad t \geq t_0,$$

or

$$v(x_0, t_0) \geq \int_{t_0}^{t_0+T} \| x_0 \|^2 \exp[-2nL(t - t_0)] \, dt = c_1 \| x_0 \|_2^2.$$

TABLE I

Properties of the Lyapunov function $v(x, t)$	Properties of the trajectory $x(x_0, t_0, t)$ of perturbed motion
1. Lyapunov's theorem on stability: the function v is positive-definite; the derivative dv/dt is negative or 0.	1. The null solution $x = 0$ is stable in the sense of Lyapunov (in the sense of definition 1.1, p. 2).
2. The function $v(x, t)$ admits an infinitely small upper bound.	2. The null solution $x = 0$ is stable uniformly with respect to t_0 (in the sense of property C, definition 9.1, p. 47).
3. The derivative dv/dt satisfies relation (10.5); i.e., $$\int^{\infty} m_\eta(\tau) = -\infty \text{ for } \eta > 0,$$ where $$m_\eta(t) = \sup dv/dt$$ for $\eta \leq \|x\|, \backslash x \in H_1$.	3. The null solution $x = 0$ is asymptotically stable. Moreover, the stability is uniform with respect to x_0 (in the sense of property D, definition 10.1, p. 54).
4. The Lyapunov theorem on asymptotic stability: the derivative dv/dt is a negative-definite function.	4. The null solution $x = 0$ is asymptotically stable, and the stability is uniform with respect to x_0, t_0 (in the sense of property B, definition 5.1, p. 26).
5. The Lyapunov function $v(x, t)$ satisfies the estimates (11.7), which are characteristic of a Lyapunov function which is a quadratic form, that is, $$c_1 \|x\|_2^2 \leq v(x, t) \leq c_2 \|x\|_2^2,$$ $$\frac{dv}{dt} \leq -c_3 \|x\|_2^2,$$ $$\left| \frac{\partial v}{\partial x_i} \right| \leq c_4 \|x\|_2,$$ for $i = 1, \cdots, n$, where $c_1, \cdots, c_4$ are positive constants.	5. The solution $x(x_0, t_0, t)$ satisfies condition (11.9); that is, $$\|x(x_0, t_0, t)\|_2$$ $$\leq B \|x_0\|_2 \exp[-\alpha(t - t_0)]$$ for $t \geq t_0$, where α, B are positive constants.

Thus the first two assertions (in the first line) of estimate (11.7) are established for the function $v(x, t)$ defined by (11.16).

(3) Differentiating, and using relations (11.11), we obtain for dv/dt

$$(11.17) \quad \frac{dv(x(x_0, t_0, t), t)}{dt} = \frac{d}{dt}\left[\int_t^{t+T} \|x(x(x_0, t_0, t), t, \tau)\|_2^2 \, d\tau \right]$$

$$= -\|x(x(x_0, t_0, t), t, t)\|_2^2 + \|x(x(x_0, t_0, t), t, t + T)\|_2^2$$

$$+ \int_t^{t+T} \frac{d}{dt}\|x(x(x_0, t_0, t), t, \tau)\|_2^2 \, d\tau .$$

By definition of the terms, we see that

$$x(x(x_0, t_0, t + \varDelta t), t + \varDelta t, \tau) = x(x(x_0, t_0, t), t, \tau) .$$

Thus, the integral in the right member of (11.17) is zero. Using the value of T and the hypothesis (11.15), we obtain

$$\| x(x(x_0, t_0, t), t, t + T) \|_2 < B e^{-\alpha T} \| x(x_0, t_0, t) \|_2 = \frac{1}{\sqrt{2}} \| x(x_0, t_0, t) \|_2 \, .$$

Thus, the derivative dv/dt must satisfy the inequality

$$\frac{dv}{dt} \leq -\| x(x_0, t_0, t) \|_2^2 + \tfrac{1}{2} \| x(x_0, t_0, t) \|_2^2 = -\tfrac{1}{2} \| x(x_0, t_0, t) \|_2^2 \, ,$$

at the point $x(x_0, t_0, t)$. Thus the second line of equality (11.7) is established for the function $v(x, t)$ of (11.16).

Now we come to the assertion of (11.7) concerning the continuity of the partial derivatives $\partial v / \partial x_i$ $(i = 1, \cdots, n)$. The fact that the derivatives in question exist and are continuous follows [39a, II, 218] from the theorem on differentiation under the integral sign in (11.16), and from the further fact that the partial derivative $\partial x_i(x_0, t_0, t)/\partial x_{j0}$ of the solution $x(x_0, t_0, t)$ is a continuous function of the initial values x_{j0} [39, theorems 7.4, 7.5, pp. 29, 30].

Thus we obtain

$$(11.18) \qquad \frac{\partial v}{\partial x_{j0}} = \int_{t_0}^{t_0+T} \sum_{i=1}^{n} 2x_i(x_0, t_0, t) \frac{\partial x_i(x_0, t_0, t)}{\partial x_{j0}} \, dt \, .$$

The derivatives $\partial x_i / \partial x_{j0}$ satisfy the inequality [138, p. 23]

$$(11.19) \qquad \left| \frac{\partial x_i(x_0, t_0, t)}{\partial x_{j0}} \right| \leq n \exp[nL \, | t - t_0 |] \quad (i = 1, \cdots, n; \; j = 1, \cdots, n) \, .$$

Combining this with (11.18), (11.15), and (11.19), we obtain

$$\left| \frac{\partial v}{\partial x_{j0}} \right| \leq \int_{t_0}^{t_0+T} 2n^2 B \| x_0 \|_2 \exp[(nL - \alpha)(t - t_0)] \, dt = c_4 \| x_0 \|_2 \, .$$

Thus the last of the estimates (11.17) for the functions $v(x, t)$ of (11.16) is established.

Now we show that the function $v(x, t)$ of (11.16) has a continuous partial derivative $\partial v / \partial t$. We compute

$$\frac{1}{\Delta t}(v(x_0, t_0 + \Delta t) - v(x_0, t_0)) = \frac{1}{\Delta t} \Big(v\big(x(x_0, t_0, t_0 + \Delta t), t_0 + \Delta t\big)$$

$$- v(x_0, t_0) - v\big(x(x_0, t_0, t_0 + \Delta t), t_0 + \Delta t\big) + v(x_0, t_0 + \Delta t) \Big) \, ,$$

so that since dv/dt and $\partial v / \partial x_i$ $(i = 1, \cdots, n)$ exist and are continuous, the relation

$$(11.20) \qquad \frac{1}{\Delta t} | v(x_0, t_0 + \Delta t) - v(x_0, t_0) | = \frac{d\tilde{v}}{dt} - \sum_{i=1}^{n} \frac{\partial \tilde{v}}{\partial x_{i0}} \tilde{X}_i$$

holds. Here, the symbol $\sim$ indicates that the corresponding derivatives are to be evaluated at an intermediate point. Substituting (11.20) into (11.19), we obtain the formula

$$\frac{\partial v}{\partial t} = \frac{dv}{dt} - \Big(X_1 \frac{\partial v}{\partial x_{10}} + \cdots + X_n \frac{\partial v}{\partial x_{n0}} \Big)$$

by allowing $\Delta t \to 0$. Thus, $\partial v / \partial t$ exists and is continuous. The theorem is proved.

Remark. Condition (11.15) is an extremely natural one for linear equations. The appropriateness of this condition for nonlinear equations will be apparent from the following considerations. It is often a fruitful approach to study a system of nonlinear equations by replacing it with some similar system. If this is cleverly done, the stability of the nonlinear system can be studied by using a connection with the auxiliary linear system in an appropriate way. See, for example, Chapter 4. The important thing is that the behavior of the trajectory of the nonlinear system should be similar to the behavior of the linear system in the case of stability. There are examples in sections 21 and 25 (Chapters 4 and 5). Thus relation (11.15) and a Lyapunov function $v(x, t)$ that satisfies estimate (11.7) are much more important in the theory of nonlinear systems than would appear at first glance.

The results already obtained in this chapter on stability and the corresponding properties of the Lyapunov function $v(x, t)$ that are necessary and sufficient for properties of the trajectory $x(x_0, t_0, t)$ are set out in Table I. The properties in each of the points from number 2 on are supplementary properties. For example, property 2 is really the logical sum of properties 1 and 2, assertion 3 is the logical sum of points 1, 2, 3, etc.

12. Modifications of Četaev's Theorem on Instability

In section 7 it was shown that whenever the null solution $x = 0$ is unstable, there is a neighborhood of this solution in which a function $v(x, t)$ exists and satisfies Četaev's theorem on instability. For certain applications it is desirable to be able to assert the existence of a function $v(x, t)$ that is subject to stronger conditions than the function v of that theorem. An important case is that in which the region $dv/dt > 0$ contains the region $v \geqq 0$ (except for the point $x = 0$). The precise formulation is given by the following definition.

DEFINITION 12.1. *The function dv/dt is said to be positive-definite in the region $v \geqq 0$ (for x in the region H_0), provided that whenever a positive number $\eta > 0$ is given, there is a number $\gamma > 0$ such that $dv/dt > \gamma$ whenever $\| x \| > \eta$, $x \in H_0$, $v(x, t) \geqq 0$.*

It is clear that if the function dv/dt satisfies the conditions of this definition, the function dv/dt is positive-definite in the region $v > 0$ in the sense of Četaev's definition [36, p. 32]. The converse is not valid in general. We now consider some modifications of Četaev's theorem.

In the modification about to be given, the function v will be subjected to stronger conditions than those of Četaev's theorem. Thus it is to be expected that the instability criterion is coarser; i.e., it cannot necessarily be used in all cases of instability. An interesting property of the function v of theorem 12.1 (below) is that the conditions satisfied by the function v of that theorem are preserved when the equations of perturbed motion are varied in a suitably restricted manner.

We consider only equations in which the right member does not depend explicitly on the time:

$$(12.1) \qquad \frac{dx_i}{dt} = X_i(x_1, \cdots, x_n) \qquad (i = 1, \cdots, n) \,,$$

where the functions X_i are defined and continuous in the region

$$(12.2) \qquad || x || < H \qquad (H = \text{const.}) \,.$$

We also require that the functions X_i have continuous partial derivatives $\partial X_i/\partial x_j$ $(i, j = 1, \cdots, n)$ in the region (12.2).

THEOREM 12.1. *Let $\bar{H}_0$ be a closed region, and suppose that for $|| x || \leqq \bar{H}_0$, the function $v(x)$ has a derivative dv/dt that is positive-definite in the region $v \geqq 0$ in the sense of definition 12.1. Suppose further that the point $x = 0$ belongs to the closure of the region $v > 0$. Then the null solution $x = 0$ is unstable, and there is a trajectory $x(x_0, t)$ that converges on the point $x = 0$ for $t \to -\infty$,*

$$(12.3) \qquad \lim || x(x_0, t) || = 0 \quad \text{for } t \to -\infty \,;$$

moreover, $|| x_0 || = H_0$.

Remark. The hypothesis does not require that the function $v(x)$ be defined in the entire region $|| x || \leqq H_0$. It is required only that the (continuous) function $v(x)$ be defined in some open region h that contains the region $v \geqq 0$ (with the possible exception of the point $x = 0$).

PROOF. Let $x_0^{(k)}$ be a sequence of points $(\neq 0)$ lying in the region $|| x || < H_0$, on the surface $v = 0$, and converging to the point $x = 0$ for $k \to \infty$. For $|| x || \leqq H_0$, $v \geqq 0$, the derivative $dv/dt > 0$ $(x \neq 0)$; hence the trajectory $x(x_0^{(k)}, t)$ moves into the region $v > 0$ for all values of $t > 0$ for which the trajectory remains in the region $x \leqq H_0$. Thus, for each $k \geqq 1$ there is a number $t_k > 0$ with the property that $|| x(x_0^{(k)}, t_k) || = H_0$, and such that $|| x(x_0^{(k)}, t) || < H_0$ for $0 \leqq t < t_k$. This is clear; if τ is any positive number for which $|| x(x_0^{(k)}, \tau) || < H_0$, the relation $v(x(x_0^{(k)}, \tau)) = \alpha_k > 0$ will hold. The function $v(x)$ is continuous, with $v(0) = 0$, because we can find a positive number δ_k which satisfies the condition

$$(12.4) \qquad | v(x) | < \alpha_k \quad \text{for } || x || < \delta_k \,.$$

Since the relation dv/dt holds along $x(x_0^{(k)}, t)$ as long as $|| x(x_0^{(k)}, t) || < H_0$, it follows that $v(x(x_0^{(k)}, t)) > \alpha_k$ for $t \geqq \tau$; thus by (12.4), $|| x(x_0^{(k)}, t) || > \delta_k$ for $t \geqq \tau$. It then follows that, since dv/dt is positive-definite in the region $v \geqq 0$, there must exist a positive constant β_k such that the relation $dv/dt > \beta_k$ holds along the trajectory $x(x_0^{(k)}, t)$ for $t > \tau$, as long as $|| x(x_0^{(k)}, t) || < H_0$. This makes it clear that the numbers t_k do indeed exist, and that, moreover, $v(x(x_0^{(k)}, t_k)) > 0$.

We consider the sequence of points $x(x_0^{(k)}, t_k)$. Since this is a bounded sequence, it must have a limit point x_0. To simplify the notation, suppose that a subsequence converging to this limit point is $x(x_0^{(k)}, t_k)$. The function $v(x)$ is continuous; thus $v(x_0) \geqq 0$. Moreover, $|| x_0 || = H_0$. We assert that the trajectory $x(x_0, t)$ approaches the point $x = 0$ for $t \to -\infty$. The first step is to show that for $t < 0$, the point $x(x_0, t)$ remains in the region $v \geq 0$, $\| x \| \leq H_0$. The proof is by contradiction. Suppose the point $x(x_0, t^*)$ is outside the region in question for $t = t^* < 0$. Since the solution depends continuously on the initial condition, it must be true that for sufficiently large k, the point $x(x_0^{(k)}, t_k + t^*)$ must lie in an arbitrarily small neighborhood of the point $x(x_0, t^*)$. Thus, it cannot lie in the region $\| x \| \leq H_0$, $v \geq 0$. However, this is impossible since the numbers t_k are so chosen that for $0 \leq t < t_k$, the relations $\| x(x_0^{(k)}, t) \| < H_0$, $v(x(x_0^{(k)}, t)) \geq 0$ must hold; moreover, $t_k \rightarrow \infty$ for $k \rightarrow \infty$. This contradiction shows that $\| x(x_0, t) \| < H_0$ and $v(x(x_0, t)) \geq 0$ for $t < 0$. Moreover, it is easily shown that $\| x(x_0, t) \| \rightarrow 0$ for $t \rightarrow -\infty$. Indeed, suppose the contrary. It would then be possible to find a monotonic sequence of values of t, $t^{(m)} < 0$, such that $t^{(m)} \rightarrow -\infty$ for $m \rightarrow \infty$, and $\| x(x_0, t^{(m)}) \| > \eta > 0$. But the speed of the point $x(x_0, t)$ when the trajectory is traversed in the phase space $(x_1, \cdots, x_n)$ is uniformly bounded. Thus there will exist a positive number ϑ $(\vartheta > 0)$ such that the relation

$$(12.5) \qquad \| x(x_0, t) \| > \frac{\eta}{2}$$

holds for $t^{(m)} - \vartheta \leq t \leq t^{(m)} + \vartheta$. Thus, the total time the point $x(x_0, t)$ spends outside the region $\| x \| < \eta/2$ must increase without limit with increasing $|t|$, because of the hypotheses. By conditions (12.5), $dv/dt > \alpha > 0$. Therefore, the assumptions show that $v(x(x_0, t)) \rightarrow -\infty$ for $t \rightarrow -\infty$. This is a contradiction, since $v(x(x_0, t)) \geq 0$ for $t < 0$. The only possibility remaining is that $v(x(x_0, t)) \rightarrow 0$ and $\| x(x_0, t) \| \rightarrow 0$ for $t \rightarrow -\infty$. This concludes the proof of the theorem.

Theorem 12.1 shows that a necessary condition for the existence of a function $v(x)$ whose derivative is negative-definite in the region $x \geq 0$ is that there be a trajectory $x(x_0, t)$ satisfying condition (12.3). It can be shown further that this condition is also a sufficient one for the existence of a function $v(x)$ with these properties.

THEOREM 12.2. *If there is a trajectory $x(x_0, t)$ that satisfies condition (12.2), $| x_0 | > H_0$, $H_0 < H$, then a function $v(x)$ can be defined that has a derivative dv/dt [evaluated along the equations (12.1)] that is positive-definite in the region $v(x) > 0$ (in the region H_0) in the sense of definition 12.1. The function $v(x)$ is defined in some region h that includes the region $v \geq 0$ (with the exception of the point $x = 0$). Moreover, the function v has continuous partial derivatives $\partial v/\partial x_i$ in this region.*

The proof of this assertion is omitted. The level surface $v = 0$ of the function $v(x)$ that satisfies the conditions of the theorem proved here is a section in the sense of Persidskiĭ [145], where the surface is actually considered in the $(n + 1)$-dimensional space $x \times t$. Therefore, we can see that if the right members X_i of equations (12.1) do not depend explicitly on the time, the condition for the existence of such a section is that there should be a trajectory $x(x_0, t)$ that converges to the point $x = 0$ as $t \rightarrow -\infty$.

3

Some Generalizations of Lyapunov's Theorems

13. Preliminary Remarks

In this chapter we give some sufficient conditions for stability or instability, based on the ideas of Lyapunov's second method. These criteria are generalizations of Lyapunov's theorems. Although the modifications of the Lyapunov theorems discussed in Chapter 2 were concerned mainly with the connection between the properties of the function v and the behavior of the trajectories $x(x_0, t_0, t)$, the generalizations of Lyapunov's theorems considered in this chapter have a different purpose. To solve concrete problems it is not always convenient to construct a function v that satisfies all the conditions of the pertinent theorem of Lyapunov. The theorems we prove below are designed to avoid certain difficulties encountered in applying Lyapunov's method to specific problems of stability. It is too much to expect that these theorems would have as universal application as Lyapunov's theorem. Also, the converses of the theorems we prove would have little value, so we do not consider them. The criteria for stability or instability derived in this chapter are applied to concrete problems of stability for nonlinear systems in Chapters 4 and 5.

14. Criteria for Asymptotic Stability

To solve the stability problem, we sometimes construct a function $v(x, t)$ that (a) is positive-definite, (b) admits an infinitely small upper bound, and (c) has a derivative dv/dt that is not negative-definite but only semidefinite in the region in question. Therefore, Lyapunov's theorem does not apply. Theorem 14.1, which follows, does apply and enables us to consider a Lyapunov function of this type in case of asymptotic stability. This theorem was proved by Barbašin and the author [16] for the case of stability in the large, when the right members X_i of the equations of perturbed motion do not depend explicitly on the time. Subsequently, Tuzov proved a similar theorem [179]. Here we consider the more general case in which the equations

$$(14.1) \qquad \frac{dx_i}{dt} = X_i(x_1, \cdots, x_n, t) \qquad (i = 1, \cdots, n)$$

of perturbed motion are such that the right members $X_i(x, t)$ are periodic functions of the time t with period ϑ, or do not depend explicitly on the time t. We further assume that these functions are defined and continuous in the region

(14.2) $\|x\| < H, -\infty < t < \infty$ ($H = \text{const.}$ or $H = \infty$)

and that the functions X_i satisfy a Lipschitz condition with respect to the variables x_j in every region $\|x\| < H_\mu < H$,

(14.3) $|X_i(x'', t) - X_i(x', t)| < L_\mu \|x'' - x'\|$ ($L_\mu = \text{const.}$).

THEOREM 14.1. *Suppose the equations of perturbed motion* (14.1) *enjoy the properties that*
(i) *there exists a function* $v(x, t)$ *which is periodic in the time t with period ϑ or does not depend explicitly on the time;*
(ii) $v(x, t)$ *is positive-definite;*
(iii) $v(x, t)$ *admits an infinitely small upper bound in the region* (14.2);
(iv)

(14.4) $\sup (v \text{ in the region } \|x\| \leq H_0, \ 0 \leq t < \vartheta)$
          $< \inf (v \text{ for } \|x\| = H_1)$ $(H_0 < H_1 < H)$;

(v)

(14.5) $dv/dt \leq 0 \text{ in the region } (14.2)$;

(vi) *the set M of points at which the derivative dv/dt is zero contains no nontrivial half-trajectory*

(14.6) $x(x_0, t_0, t)$ $(0 < t < \infty)$

of the system (14.1).
Under these conditions, the null solution $x = 0$ is asymptotically stable and the region $\|x\| \leq H_0$ lies in the region of attraction of the point $x = 0$.

PROOF. The Lyapunov stability of the solution $x = 0$ is proved, and the inequality $\|x(x_0, t_0, t)\| < H_1$ for $\|x_0\| < H_0$, $t > t_0$ is established as on p. 19 of this book, since the hypotheses of theorem 14.1 are sufficient for these assertions. Thus, we need only prove further that whenever $\|x_0\| \leq H_0$ (that is, whenever $\|x(x_0, t_0, t)\| < H_1$) there cannot be a trajectory along which $\|x(x_0, t_0, t)\| > \eta > 0$ for all $t \geq t_0$. Let us suppose the opposite—that such a trajectory exists. The function $v(x(x_0, t_0, t), t)$ is a monotonic nonincreasing function of the time t, since the inequality $dv/dt < 0$ is valid in the region (14.2). Thus the limit $\lim v(x(x_0, t_0, t), t) = v_0$ exists for $t \to \infty$. Moreover, one clearly has the relation

(14.7) $v(x(x_0, t_0, t), t) \geq v_0$

in the range $t_0 \leq t < \infty$.
If the functions $X_i(x, t)$ are periodic in the time t, let ϑ be the period; if the functions X_i do not depend explicitly on the time, let ϑ be an arbitrary positive number. Let us consider the sequence of points $x^{(k)} = x(x_0, t_0, t_0 + k\vartheta)$

$(k = 1, 2, \cdots)$. Since this sequence is bounded, it has a limit point $x = x_0^* \neq 0$. By a change of notation, suppose that $x^{(k)}$ is a convergent subsequence, convergent to the point x_0^*. Since the function $v(x, t)$ is continuous and periodic, the relation $v_0 = v(x_0^*, t_0)$ must hold. Let us consider the trajectory $x(x_0^*, t_0, t)$ for $t_0 \leq t < \infty$. By hypothesis (vi) of the theorem this half-trajectory does not lie entirely in the region M. Hence the half-trajectory must contain points for which $dv(x(x_0^*, t_0, t), t)/dt < 0$. Therefore, there is a moment of time $t^* < t_0$ for which the relation $v(x(x_0^*, t_0, t^*), t^*) = v_1 < v_0$ holds. The solution of the differential equations depends continuously on the initial conditions; since the sequence $x_0^{(k)}$ converges to the point x_0^*, the inequality

$$(14.8) \qquad \| x(x_0^*, t_0, t^*) - x(x_0^{(k)}, t_0, t^*) \| < \gamma$$

holds for all sufficiently large k, $k > N(\gamma)$. Therefore, we have

$$(14.9) \qquad \lim_{k \to \infty} v(x(x_0^{(k)}, t_0, t^*), t^*) \leq v_1.$$

The substitutions

$$x(x_0^{(k)}, t_0, t^*) = x(x_0, t_0, t^* + k\vartheta), \quad v(x, t^*) = v(x, t^* + k\vartheta)$$

follow from the periodicity of the function X_i and the function $v(x, t)$. Thus the relation (14.9) can be rewritten as follows:

$$(14.10) \qquad \lim_{k \to \infty} v(x(x_0, t_0, k\vartheta + t^*), k\vartheta + t^*) \leq v_1.$$

Since inequality (14.10) contradicts inequality (14.7), the theorem is proved. Indeed, $v_1 < v_0$.

Remark. Suppose the set of points M of condition (vi) of the theorem happens to be a surface defined by the equation

$$(14.11) \qquad F(x_1, \cdots, x_n, t) = 0.$$

Then condition (vi) will always be satisfied if the inequality

$$(14.12) \qquad \sum_{i=1}^{n} \frac{\partial F}{\partial x_i} X_i + \frac{\partial F}{\partial t} \neq 0$$

holds on the surface $F(x, t) = 0$ (in the region $\| x \| < H_1$). Indeed, if the point $x(x_0, t_0, t)$ hits the surface (14.11) at some moment of time $t = t_1$, then it cannot remain on the surface for $t > t_1$, since

$$\left[\frac{dF(x(t), t)}{dt} \right]_{t=t_1} = \sum_{i=1}^{n} \frac{\partial F}{\partial x_i} X_i + \frac{\partial F}{\partial t} \neq 0$$

in view of relation (14.12).

15. Sufficient Criteria for Instability

In this section we establish a generalization of Lyapunov's first theorem on stability for equations of perturbed motion in which the functions X_i are periodic. This generalization is related to Lyapunov's theorem 2 [114,

p. 261] in the same way that theorem 14.1 of the preceding section is related to Lyapunov's theorem on asymptotic stability.   That is, under certain circumstances, we can assume that the derivative dv/dt is semidefinite instead of definite.   The theorem in question was established by the author [91] for the case in which the right members X_i of equations (1.3) do not depend explicitly on the time.   Conditions for the existence of a function v which come up in the present section are not actually more general than the condition for the existence of a function v in Lyapunov's second theorem. In this respect, the theorem proved below is less strong than the universal theorem on instability (Lyapunov's theorem 3 [114, p. 266], and Četaev [36, p. 34]), which is always applicable.   Nevertheless, the theorem is of use in some concrete cases.   Examples of the application of these theorems to certain problems will be presented in Chapter 5.

We shall assume that the equations of perturbed motion satisfy all the conditions imposed in section 14.

THEOREM 15.1.   *Let $H_1 < H$.   Suppose that there exists a function $v(x, t)$, which is periodic in the time or does not depend explicitly on the time, such that*

(i)   *v is defined in the region (14.2);*

(ii)   *v admits an infinitely small upper bound in this region;*

(iii)

(15.1) $$\frac{dv}{dt} \geq 0 \text{ in the region (14.2), along a trajectory of (14.1) ;}$$

(iv)   *the set of points M at which the derivative dv/dt is 0 contains no nontrivial half-trajectory*

(15.2) $$x(x_0, t_0, t)   \qquad (t_0 \leq t < \infty) .$$

Suppose further that in every neighborhood of the point $x = 0$, there is a point x_0 such that for arbitrary $t_0 \geq 0$ we have $v(x_0, t_0) > 0$.   Then the null solution $x = 0$ is unstable, and the trajectories $x(x_0, t_0, t)$ for which $v(x_0, t_0) > 0$ leave the region $\| x \| < H_1$ as the time t increases.

PROOF.   Let $\delta > 0$ and $t_0 \geq 0$ be preassigned.   The hypotheses of the theorem assert that a point x_0 exists for which $\| x_0 \| < \delta$, and $v(x_0, t_0) > 0$. The assertion of the theorem is that there is a value $t_1 > t_0$ for which $\| x(x_0, t_0, t_1) \| = H_1$.

Suppose the theorem is not true.   Then, for all $t \in [0, \infty)$, the relation $\| x(x_0, t_0, t) \| < H_1$ holds.   The function $v(x, t)$ admits an infinitely small upper bound, so that there is a number $\eta > 0$ for which

(15.3) $$|v(x, t)| < v(x_0, t_0),   \quad \| x \| < \eta,   \quad 0 \leq t < \infty .$$

Since the relation $dv/dt \geq 0$ holds in the region $\| x \| < H$, our assertion, in conjunction with (15.3), is sufficient to establish the relation $\| x(x_0, t_0, t) \geq \eta$ for $t \in (t_0, \infty)$.   Thus, for $t \in [t_0, \infty)$ the relation

(15.4) $$\eta \leq \| x(x_0, t_0, t) \| < H_1$$

can be established by the use of condition (16.2), arguing by contradiction exactly as was done in the proof of theorem 14.1. The details are not included. This certainly contradicts inequality (15.3), and thus the trajectory $x(x_0, t_0, t)$ must not remain indefinitely in the region $||x|| < H_1$. The theorem is proved.

The remark at the end of the preceding section is also valid in the present case. Thus if the set M is defined by an equation (14.11), then condition (iv) of theorem 15.1 follows from inequality (14.12).

16. Criteria for Stability with Respect to Large Initial Perturbations, Based on Lyapunov's Theory of Stability of the Perturbed Trajectory $x(x_0, t_0, t)$

In this section we consider the connection between asymptotic stability of the null solution $x = 0$ of the system

$$(16.1) \qquad \frac{dx_i}{dt} = X_i(x_1, \cdots, x_n, t), \quad i = 1, \cdots, n$$

for large initial deviations x_{j0} and the stability of the trajectories $x(x_0, t_0, t)$ under sufficiently small perturbations $z_0 = x - x(x_0, t_0, t_0)$.

Here we shall assume that the right members of equation (16.1) are defined in the region

$$(16.2) \qquad 0 \leq t < \infty, \quad -\infty < x_i < \infty, \quad i = 1, \cdots, n \,,$$

and satisfy the Lipschitz condition

$$(16.3) \qquad |X_i(x'', t) - X_i(x', t)| \leq L_\mu ||x'' - x'||$$

in every bounded region

$$(16.4) \qquad ||x|| < H_\mu \,.$$

First, we prove a simple preliminary proposition.

LEMMA 16.1. *Let the null solution $x = 0$ of equation (16.1) be stable uniformly with respect to the time t_0, in the sense of property* C *(definition 9.1, p. 47). A necessary and sufficient condition that the unperturbed motion $x = 0$ be asymptotically stable and that the region $||x|| < H_0$ $(H_0 < H)$ lie in the region of attraction of the point $x = 0$ is that every perturbed motion $x(x_0, t_0, t)$ $(||x_0|| < H_0)$ be asymptotically stable for all sufficiently small initial perturbations $z_0 = x - x(x_0, t_0, t_0)$, uniformly with respect to the coordinate z_0, in the sense of property* D *(definition 10.1, p. 54).*

PROOF. First we show that the conditions of the lemma are sufficient. Let $\bar{g}_0$ be a bounded closed region, $\bar{g}_0 \subset (||x|| < H_0)$, and let $x_0 \in \bar{g}_0$. We consider the point x_0 to be contained in a neighborhood

$$(16.5) \qquad ||x_0' - x_0|| < \beta(x_0), \ \beta(x_0) > 0$$

and suppose that

$$(16.6) \qquad ||x(x_0', t_0, t) - x(x_0, t_0, t)|| < \varepsilon \quad \text{for all} \quad t \leq t_0$$

where ε is an arbitrary preassigned positive constant, and

$$(16.7) \qquad \lim_{t \to \infty} ||x(x_0', t_0, t) - x(x_0, t_0, t)|| = 0$$

for all points $x = x_0'$ in the neighborhood (16.5).   Conditions (16.5), (16.6), and (16.7) are no restriction on the choice of x_0 if $\bar{g}_0$ satisfies the conditions stated at the beginning of the proof.

The bounded closed set $\bar{g}_0$ may be covered by a finite number of neighborhoods (16.5), say by N neighborhoods.   Inequality (16.6) shows that for each fixed value $t \geq t_0$, the diameter of the region consisting of those points $x(x_0, t_0, t)$ for which $x_0 \in \bar{g}_0$ is bounded by the number $2\varepsilon N$, and, moreover, approaches 0 as $t \to \infty$ because of the uniformity of the stability of $x(x_0, t_0, t)$ with respect to z_0.

Since every point $||x_0|| < H_0$ lies in at least one such region $\bar{g}_0$ of the type defined above, the asymptotic stability of the null solution $x = 0$ with respect to perturbations of x_0 within the region $||x|| < H_0$ has been established, and thus the conditions of the lemma are sufficient.

Now let us show that the conditions of the lemma are necessary.   Suppose $||x_0|| < H_0$ and let $\bar{g}_0$ be a bounded closed region contained in $||x|| < H_0$, and suppose further that $x_0 \in \bar{g}_0$.   Under the hypotheses of the lemma, the null solution $x = 0$ is asymptotically stable with respect to the coordinate x_0 in the region g_0 (cf. definition 10.1, lemma 10.1, p. 54).   Thus, for every preassigned positive number ε, there exists a number $T(\varepsilon/2, t_0)$ such that the relation

$$(16.8) \qquad\qquad ||x(x_0', t_0, t)|| < \frac{\varepsilon}{2}$$

holds for all initial values $x_0' \in \bar{g}_0$, for all $t \geq t_0 + T(\varepsilon/2, t_0)$.   Since the solution $x(x_0', t_0, t)$ depends continuously on the initial value x_0', it will be possible to find a neighborhood (16.5) that lies entirely in the region $||x|| < H_0$ so that the inequality $||x(x_0, t_0, t) - x(x_0', t_0, t)|| < \varepsilon/2$ holds for all t for which $t_0 \leq t \leq t_0 + T(\varepsilon/2, t_0)$, and all x_0' for which (16.5) holds.   It then follows that inequality (16.6) must be satisfied for all points x_0' that lie in the region (16.5), so that the solution $x(x_0, t_0, t)$ is necessarily stable.

Now we show that this solution is asymptotically stable.   Let η be an arbitrary positive number.   As above, we see that there must be a number $T(\eta/2, t_0)$ such that the relation

$$||x(x_0', t_0, t)|| < \eta/2$$

holds for all $t \geq t_0 + T(\eta/2, t_0)$ and for all $x_0' \in g_0$, and, therefore, for all x_0' lying in the neighborhood (16.5) of the point x_0.   Thus it is clear that the inequality

$$||x(x_0', t_0, t) - x(x_0, t_0, t)|| < \eta$$

is satisfied for $t \geq t_0 + T(\eta/2, t_0)$.   Thus the solution $x(x_0, t_0, t)$ is indeed asymptotically stable uniformly with respect to the coordinate

$$z_0 = x_0' - x(x_0, t_0, t)$$

for $||z_0|| < \beta(x_0)$, where $\beta(x_0)$ is the constant of inequality (16.5).   The lemma is proved.

Remark.   If we repeat the considerations used in the proof of the lemma, but assume further that the asymptotic stability is uniform with respect to the coordinates $x_0 \in \bar{g}_0$ $(\bar{g}_0 \in (\|x\| < H_0))$ and also with respect to the time t_0, in the sense of property B (definition 5.1, p. 26), we conclude that a necessary and sufficient condition for stability of this kind is that the solution $x(x_0, t_0, t)$ be asymptotically stable for sufficiently small perturbations $z_0 = x - x_0$. This must be true uniformly with respect to z_0 and t_0 in the sense of property B. Moreover, for each positive $\eta > 0$, the value of the number $T(\eta)$ that appears in property B must be uniformly applicable to all the trajectories $x(x, t_0, t)$ for which $x_0 \in g_0$.

Now we shall consider asymptotic stability of the null solution $x = 0$ for arbitrary initial perturbations, $-\infty < x_{j0} < \infty$. We assume that the null solution $x = 0$ is asymptotically stable in the sense of property B uniformly in t_0 and $x_0 \in \bar{g}_0$, for every bounded, closed region $\bar{g}_0$. From lemma 16.1 (and the remark following the lemma), the theorems of Malkin [125], the theorem in references [16] and [97], and theorem 5.3 of this book have the following consequence.

THEOREM 16.1.   *Suppose that a function $v(x, t)$ is defined in the entire space $-\infty < x_i < \infty$, $0 \leq t < \infty$, and has the following properties:*
(i)   *$v(x, t)$ is positive-definite;*
(ii)   *$v(x, t)$ admits an infinitely small upper bound;*
(iii)   *$v(x, t)$ admits an infinitely great lower bound;**
(iv)   *the derivative dv/dt is negative-definite for all x and $t \geq 0$.   Then in the neighborhood*

$$\|x(x_0, t_0, t) - x\| < \beta(x_0), \qquad \beta(x_0) > 0,$$

every trajectory $x(x_0, t_0, t)$ possesses a positive-definite function

$$v(x_0, t_0, z, t), \qquad z = x - x(x_0, t_0, t),$$

and this last function admits an infinitely small upper bound and has a derivative that is a negative-definite function along the trajectories that satisfy

$$\frac{dz_i}{dt} = Z_i(z_1, \cdots, z_n, t),$$

(16.9) $$Z_i(z, t) = X_i(x, t) - X_i(x(x_0, t_0, t), t),$$

$$z_i = x_i - x_i(x_0, t_0, t), \qquad i = 1, \cdots, n.$$

In describing $v(x_0, t_0, z, t)$ as positive-definite, the variable argument is to be $z(t)$.   Conversely, if the latter function exists in every neighborhood of the above type, then there will be a function $v(x, t)$ having properties (i)–(iv).   When a function $v(x_0, t_0, z, t)$ exists with the above properties, the positive-definiteness, the infinitely small upper bound, and the negative-definiteness of the derivative will be uniform with respect to $x_0 \in \bar{g}_0$, $t_0 \in [0, \infty)$.   In other words, in every region $\bar{g}_0$, functions $w_1(z), w_2(z), W(z)$ can be found that have the properties

---

* See p. 29.

$$W(0) = 0, \quad w_1(z) > 0, \quad w_2(z) > 0 \quad for \quad z \neq 0$$

and

(16.10)
$$w_1(z) \leqq v(x_0, t_0, z, t) \leqq W(z),$$
$$\frac{dv}{dt} = -w_2(z) \quad for \quad x_0 \in \bar{g}_0.$$

Theorem 16.1 and lemma 16.1 reduce the problem of stability for large initial perturbations to the problem of stability for small perturbations. Now this last problem can be linearized except in critical cases. That is, the problem can be solved by studying a truly linear system of equations. For this it is important to have information about Lyapunov functions that are quadratic forms. This is the subject of the considerations below. It is especially important to specialize the results of theorem 16.1 under the further assumption that the function $v(x_0, t_0, z, t)$ satisfies estimates (11.7), which characterize Lyapunov functions that are quadratic forms. To establish criteria for asymptotic stability based on functions $v(x_0, t_0, z, t)$ of this kind is the problem discussed here. We note that theorem 16.1 presupposed that the neighborhood $|| x - x(x_0, t_0, t) || < \delta$ of the perturbed trajectory $x(x_0, t_0, t)$, in which the function $v(x_0, t_0, z, t)$ is defined and in which it satisfies the conditions of Lyapunov's theorem, does not depend explicitly on the time t. In this respect the hypothesis is similar to Lyapunov's theorem on stability. Moreover, in particular, the region in question does not shrink to a point if the time increases indefinitely. In particular applications, it might be true that the proof of the existence of a Lyapunov function v in a neighborhood of a trajectory of this kind could be carried through only if the neighborhood is known not to shrink to a point when the time t increases. To prove that this last does not occur may be especially difficult.

Thus we find it desirable to prove another theorem that circumvents this difficulty in a number of cases. In Chapter 5 we shall derive a concrete criterion for asymptotic stability based on the theorem that follows.

THEOREM 16.2. *Let $\bar{h}_0$ be a bounded, closed, connected region; suppose that for every $x_0 \in \bar{h}_0$ there exists a function $v(x_0, t_0, z, t), z = x - x(x_0, t_0, t)$ that satisfies the following inequalities*

(16.11)
$$c_1 || z ||_2^2 \leqq v(x_0, t_0, z, t) \leqq c_2 || z_2 ||_2^2,$$

(16.12)
$$\left(\frac{dv(x_0, t_0, z(t), t)}{dt} \right)_{\text{along (16.9)}} \leqq -c_3 || x ||_2^2,$$

where c_1, c_2, c_3 are positive constants whose values do not depend on the point x_0. Moreover, the inequalities are supposed to be valid for all x that satisfy

(16.13)
$$|| x - x(x_0, t_0, t) ||_2 < \delta(x_0, t_0, t),$$

where $\delta(x_0, t_0, t)$ is a function with the following properties:
  (i) $\delta(x_0, t_0, t) > 0, \ t \geqq t_0;$

(ii) *for every $t_1 > t_0$, there is a number $\gamma(t_1) > 0$ such that $\delta(x_0, t_0, t) > \gamma(t_1)$ for all t, $t_0 \leqq t \leqq t_1$, $x_0 \in \bar{h}_0$.*

Then the null solution $x = 0$ is asymptotically stable and the region h_0 lies in the region of attraction of the point $x = 0$.

Remark. A sufficient condition that the function $\delta(x_0, t_0, t)$ satisfy the conditions of the theorem is that it be positive and continuous for $t \geqq t_0$, $x_0 \in \bar{h}_0$. Indeed, the mere continuity of the function $\delta(x_0, t_0, t)$ in the region $t_0 \leqq t \leqq t_1$, $x_0 \in \bar{h}_0$ is enough to assure that it has a positive minimum m, which can be used to define $\gamma(t_1)$ by the relation $m = 2\gamma(t_1)$.

PROOF. Let q be a number in the interval 0, 1: $0 < q < 1$, and let T be defined by $c_1 c_3 T = \log c_2 - \log (c_1 q^2)$. By hypothesis, a function $v(x_0, t_0, z, t)$ is defined in the region

$$t_0 \leqq t < t_0 + T, \quad || x - x(x_0, t_0, t) ||_2 \leqq \gamma(t_0 + T)$$

and satisfies estimates (16.11) and (16.12) there. We note that if

$$\eta = \frac{\gamma}{2} \left(\frac{c_1}{c_2} \right)^{1/2},$$

the value of $v(x_0, t_0, z, t)$ for $|| z ||_2 \leqq \eta$ is smaller than any value attained by the function v for $|| z ||_2 = \gamma$, $t_0 \leqq t \leqq t_0 + T$. Since the relation $dv/dt \leqq 0$ holds in the region (16.13), it follows that the trajectory $x(x_0', t_0, t)$ remains in the region $|| x - x(x_0, t_0, t) ||_2 \leqq \gamma$ when x_0' satisfies $|| x_0' - x_0 ||_2 < \eta$, $t_0 \leqq t \leqq t_0 + T$. Thus the trajectory remains *a fortiori* in the region (16.13), where the relations (16.11) and (16.12) are satisfied. Thus the inequality (16.12) can be integrated on $(t_0, t_0 + T)$ with respect to t, as long as $|| x_0' - x_0 ||_2 < \eta$. In this integration, we take account of estimate (16.11), recall the choice of the number T, and obtain the estimate

(16.14) $|| x(x_0', t_0, t_0 + T) - x(x_0, t_0, t_0 + T) ||_2 \leqq q || x_0 - x_0' ||_2$.

This inequality shows that as t increases from t_0 to $t_0 + T$, the diameter of an arbitrary sphere

(16.15) $|| x - x_0 ||_2 \leqq R^2$

that is entirely contained in the region h_0 is diminished in the ratio q. This is true because on every radius of such a sphere there is a finite number of points $x_0, x_0', \cdots$ such that any two of them satisfy the inequality

(16.16) $|| x_0' - x_0 || < \eta$.

Since the distance between every two such points is diminished as in (16.14), it is clear that the distance between the center x_0 of the sphere and any point on its surface is also diminished in the ratio q. We note also that for any t, $t_0 \leqq t \leqq t_0 + T$, the relation

(16.17) $|| x(x_0', t_0, t) - x(x_0, t_0, t) ||_2 < \left(\frac{c_2}{c_1} \right)^{1/2} R$

holds, provided x_0, x_0' both lie interior to the sphere (16.15).

If two points x_0'', x_0' lie on a radius of the sphere (16.15) and satisfy inequality (16.16), then the relations

$$\text{(16.18)} \qquad \frac{dv}{dt} \leq 0, \quad t_0 \leq t \leq t_0 + T \, ,$$

$$\text{(16.19)} \qquad v(x_0', t_0, x_0'' - x_0', t_0) \leq c_2 \| x_0'' - x_0' \|_2^2 \, ,$$

$$\text{(16.20)} \quad v(x_0', t_0, x(x_0'', t_0, t) - x(x_0', t_0, t), t) \geq c_1 \| x(x_0'', t_0, t) - x(x_0', t_0, t) \|_2^2$$

hold. Relation (16.17) is easily established by combining these relations in the obvious way.

Now let the interval of time T increase without bound. It is clear that the number q approaches 0, and from this we see that every solution $x(x_0, t_0, t)$ is asymptotically stable with respect to perturbations $z_0 = x - x(x_0, t_0, t)$ lying in the sphere (16.15). Thus lemma 16.1 shows that the null solution $x = 0$ is asymptotically stable with respect to perturbations x_0 lying in the region h_0. The theorem is established.

Now we suppose that the region h_0 in which the conditions of theorem 16.2 are satisfied contains the entire space $-\infty < x_i < \infty$ ($i = 1, \cdots, n$), and that the numbers c_1, c_2, c_3, which appear in estimates (16.11) and (16.12), are uniformly applicable for $0 \leq t < \infty$. It is then clear that the null solution $x = 0$ will be stable with respect to arbitrary changes $-\infty < x_{j0} < \infty$ ($j = 1, \cdots, n$). Moreover, the trajectory $x(x_0, t_0, t)$ will satisfy the inequality

$$\text{(16.21)} \qquad \| x(x_0, t_0, t) \|_2 \leq B \| x_0 \|_2 \exp \left[-\alpha(t - t_0) \right]$$

for all initial perturbations x_0 in the case at hand. Here B and α are positive constants. Inequality (16.17) shows that the value of B can be taken as $B = (c_1/c_2)^{1/2}$. From the choice of the number T, inequality (16.14) shows that the number α can be taken to be $\alpha = 1/(2c_1c_3)$. Thus theorem 11.1 is enough to yield the truth of the assertion in question.[*]

THEOREM 16.3. *Suppose that a function* $v(x_0, t_0, z, t)$ *exists and satisfies estimates* (16.11) *and* (16.12) *in a neighborhood of every trajectory* $x(x_0, t_0, t)$, $\| x_0 \|_2 < \infty$. *Then a function* $v(x, t)$ *can be defined in the entire space* $\| x \|_2 < \infty$, $0 < t < \infty$, *and the function* $v(x, t)$ *satisfies estimate* (11.7). *This means that the relations*

$$\text{(16.22)} \qquad c_1' \| x \|_2^2 \leq v(x, t) \leq c_2' \| x \|_2^2 \quad \textit{for all} \quad x \, ,$$

$$\text{(16.23)} \qquad \frac{dv}{dt} \leq -c_3 \| x \|_2^2 \quad \textit{for all} \quad x \, ,$$

$$\text{(16.24)} \qquad \left| \frac{\partial v}{\partial x_i} \right| < c_4'(R) \| x \|_2 \quad \textit{for} \quad \| x \|_2 < R$$

hold ($c_1', c_2', c_3', c_4' > 0$ *are positive constants*).

---

[*] Theorem 11.1 was proved in the case where the functions X_i are continuously differentiable. By a more complicated proof it is possible to establish the validity of the theorem when the differentiability is replaced merely by the Lipschitz condition (1.7).

17. Some Complements to the Theorems of Lyapunov

The application of Lyapunov's theorems is as follows. In studying a system of equations (1.3) by the methods of one of Lyapunov's theorems (I, II, or III [114, pp. 259–66]), we first construct a Lyapunov function v. Next, we check that the function v satisfies all the conditions of the theorem in question. Last of all, we examine the derivative dv/dt to see that it has the necessary properties on all the trajectories of (1.3) in the region H.

It is clearly possible to think of generalizing these theorems in various directions.

Let us consider a particular trajectory $x(x_0, t_0, t)$ and think of constructing a function v that satisfies the conditions of a Lyapunov theorem in a region H, with the exception of the properties that the derivative dv/dt is required to have; for this, we suppose only that the conditions of Lyapunov are satisfied along the particular trajectory $x(x_0, t_0, t)$ under consideration in the region H. Now if such a function v exists for every trajectory $x(x_0, t_0, t)$, and if certain necessary uniformity properties of v are satisfied for all trajectories in the region H, then the conclusions of Lyapunov's theorem remain valid. It should be remarked that this generalization of Lyapunov's theorem has one objection: a proof based on this theorem would have to be obtained by the method of contradiction.

Here we give such a generalization of Lyapunov's theorem in the case of asymptotic stability. (This formulation is used later, in section 24, pp. 104–7. An attempt to generalize the other Lyapunov theorems in the same way would be unsuccessful.) We shall assume that the equations of perturbed motion are defined in the region $||x|| < H$.

THEOREM 17.1. *Let* $H_0 < H_1 < H$. *Let* x_0 *be such that* $||x_0|| < H_0$. *Suppose that for every* x_0 *in this region and for every positive* t_0 *there is a function* $v(x_0, t_0, x, t)$ *that is defined in the region* $||x|| < H$, $t_0 < t < \infty$, *and that has the following properties*:

(i) $v(x_0, t_0, x, t)$ *is positive-definite in the region* $||x|| < H$, $t_0 < t < \infty$.

(ii) $v(x_0, t_0, x, t)$ *admits an infinitely small upper bound in the region* H. *Moreover, this bound is uniform with respect to* x_0, t_0; *that is,*

(17.1) $$w(x) \leqq v(x_0, t_0, x, t) \leqq W(x) \quad for \quad ||x|| < H .$$

Here $w(x)$, $W(x)$ *are continuous functions which are independent of* x_0, t_0 *and for which* $w(x) > 0$, $x \neq 0$, $W(0) = 0$, *and*

(17.2) $$\sup_{||x||=H_0} w(x) < \inf_{||x||=H_1} W(x) .$$

(iii) *The derivative* dv/dt *is negative-definite along a trajectory* $x(x_0, t_0, t)$ *of the equations of perturbed motion; that is,*

(17.3) $$\frac{dv(x_0, t_0, x(x_0, t_0, t), t)}{dt} \leqq -w_1(x_0, t_0, x)$$

in the points $||x(x_0, t_0, t)|| < H_1$, $x \neq 0$, *where*

(17.4) $$w_1(x_0, t_0, x) > 0, \quad x \neq 0 .$$

Then the null solution $x_1 = \cdots = x_n$ is asymptotically stable, and the region $\| x \| < H_0$ lies in the region of attraction of the point $x = 0$.

Remark. If the function $w_1(x_0, t_0, x)$ of relations (17.4) and (17.3) does not depend explicitly on x_0, t_0, then the solution $x = 0$ of theorem 17.1 will be asymptotically stable uniformly in x_0, t_0 (in the sense of property B, definition 5.1, p. 26).

We now turn to the question of extending Lyapunov's methods to the case that the right members of the equations of perturbed motion are discontinuous functions. Such problems do arise in a number of applications. (See [109], [113], where servomechanisms are shown to give rise to such equations.) In [178] it appears that discontinuous functions give the solution of an optimal problem; the corresponding system of equations is singular.

Where the functions X_i are discontinuous, Lyapunov's method has been used to obtain interesting and important results. See Lur'e [113], Letov [108], [109], [110], Aizerman [5].* The desirability of extending Lyapunov's method to equations with a discontinuous right member is clear. It is just as clear that some modification is necessary in the original proof of Lyapunov [114] or in the proofs given in other sources [36], [40], [124], where the functions X_i are assumed to be continuous. A detailed proof of theorem 17.1 can be given by modifying Lyapunov's proof in an appropriate way [114, pp. 259–60].

The considerations below cannot presume to be rigorous since the theory of differential equations in which the right members X_i are discontinuous has never been completely worked out.**

In [24] Pontryagin and Boltyanskiĭ took the obvious approach of carrying Lyapunov's method over to the case under present consideration by writing the functions X_i as limits of continuous functions $\tilde{X}_i$. However, a general theory for equations with discontinuous right members X_i is in such a state that in a large number of cases it is not clear how to make a rigorous definition for trajectories on a surface, even when the discontinuous functions X_i are linear (for example, in the so-called sliding regime, or regime of sliding). It will be sufficient to give a simple solution of the question in case the surface of discontinuity of the functions X_i is a smooth section for the trajectories of the corresponding system of equations.

We will say that a neighborhood of the point $x = x_0$ (in the space $x \times t$) on a surface S in the product space $x \times t$ is a trajectory of smooth section when the functions X_i are discontinuous, if the surface S is defined in the $(n + 1)$-dimensional space $x \times t$ by an equation

(17.5) $$F(x_1, \cdots, x_n, t) = 0 ,$$

where the function F has continuous partial derivatives $\partial F/\partial x_i$, $\partial F/\partial t$ in some

---

* Aĭzerman and Gantmaher give an interesting generalization of Lyapunov's theorem on first-order stability, for systems with discontinuous right members [6].

** There is a development of this theory in Filippov [58].

neighborhood of the point x_0, t_0, and where, moreover, the condition

$$(17.6) \qquad \sum_{i=1}^{n} \frac{\partial F}{\partial x_i} X_i + \frac{\partial F}{\partial t} > 0$$

is satisfied at every point in this neighborhood, or perhaps only in those points of the neighborhood which do not lie on the surface S.

Condition (17.6) states that the relation $dF(x(t), t)/dt > 0$ is valid along trajectories of the system in the two portions of space $F < 0, F > 0$. Thus in the portion of space characterized by the relation $F > 0$, the trajectories leave the surface $F = 0$, and in the portion of space characterized by the relation $F < 0$, the trajectories impinge on the surface $F = 0$ as time increases. In this way it is clear that trajectories can be continued on the surface S ($F = 0$) in a natural way. In this continuation it appears that the function $x_i(x_0, t_0, t)$ has a right derivative dx_i/dt at each point x_0, t_0, defined by $\lim_{t \to 0+} X_i(x(x_0, t_0, t), t)$. This makes it possible to define the value of the derivative dv/dt in a point (x_0, t_0) of the surface S by means of the formula

$$\left(\frac{dv}{dt} \right)_{dt = +0} = \lim_{x \to x_0, t \to t_0} \left[\sum_{i=1}^{n} \frac{\partial v}{\partial x_i} X_i(x, t) + \frac{\partial v}{\partial t} \right],$$

where it is understood that the point (x, t) lies in the region of space $F > 0$.

An important special case is that in which the section S is a surface defined by $t = \tau = \text{constant}$. This case arises when the function F is defined by $F = t - \tau$. Thus, if the functions X_i are discontinuous with respect to the time t only, no difficulty arises in applying Lyapunov's method.

To formulate all of Lyapunov's theorems in the generalization being considered, the pertinent conditions on the derivative dv/dt of the function $v(x(t), t)$ along a trajectory $x(t)$ are applied not to the true derivative dv/dt but to the expression

$$(17.7) \qquad \lim_{\Delta t \to +0} \sup \frac{v(x(t + \Delta t), t + \Delta t) - v(x, t)}{\Delta t},$$

and the corresponding generalizations can be carried through without trouble. Thus, whenever the trajectory $x(t)$ can be defined in a natural manner from the system of equations, and when a condition (17.7) can be verified along these trajectories, Lyapunov's method admits a rigorous foundation, and no difficulties arise.

Difficulties do arise when it is unreasonable or impossible to define the trajectories $x(x_0, t_0, t)$ on a surface S of discontinuity of the functions X_i. (See Fig. 4.) Suppose, for example, that there is a function $v(x, t)$ that is defined in every neighborhood of the surface S and that has a negative-definite derivative

$$(17.8) \qquad \frac{dv}{dt} = \sum_{i=1}^{n} X_i \frac{\partial x}{\partial x_i} + \frac{\partial v}{\partial t},$$

but that the trajectories $x(t)$ approach the surface S from both sides.   In this case, it is not clear how to define a trajectory $x(t)$ of the given system on the surface S.

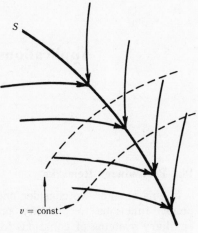

Now the equations of motion will in general be used to study a system that is in reality more complicated than that described by these equations in that there are small forces and deviations not accounted for in the equations.   Thus we find it reasonable to assume that the point $x(t)$ of the trajectory in k-space cannot remain arbitrarily long on a surface S of discontinuity if the dimension of this surface is smaller than the dimension of the entire space.   This hypothesis is a good one, since points on

FIGURE 4.

the surface S in the region where the function X_i is continuous will be subjected to random forces and displacements.   (If these random forces are sufficiently great, then the time intervals during which the point $x(t)$ remains on the surface S will constitute an inconsequentially small portion of the entire time taken by the point $x(t)$ to move along its trajectory.) However, the value of the derivative dv/dt is negative-definite throughout the region where the functions X_i are continuous.   We can therefore claim that Lyapunov's second method remains valid,[*] provided that (a) the functions X_i are discontinuous on a surface of discontinuity S (which is an $(n, n-1, \cdots)$-dimensional surface of simple shape), and (b) the conditions of the theorem in question are respectively satisfied at all points outside this special surface.

The practical question of stability of guidance and control systems can only be studied rigorously if the functions X_i appearing in the right members of the equation are allowed to include an impulsive term of the type $\delta(t)$ that is a function of the time.   We do not discuss the fundamental question of the existence of trajectories $x(x_0, t_0, t)$ of such differential equations.   Rather, we take as a hypothesis the existence of functions $x(x_0, t_0, t)$ that are defined for $t > 0$ and are solutions of the equations in some sense. Further details are given in [98].   If we substitute into a Lyapunov function a function $x(t)$ that is a known solution of the equation, we obtain a function $V(t) = v(x(t), t)$, a function of the time alone.   Conditions sufficient for asymptotic stability or instability (uniform or nonuniform) may then be imposed on the function dv/dt, but we need not require that the latter be continuous; rather, we replace dv/dt formally by the value (17.7).   The details of such a generalization of Lyapunov's theorems are omitted.

---

[*] See p. 77 for another approach to the same question.

4

Applications of Lyapunov's Method to Some General Problems of Stability

18. Preliminary Remarks

In this chapter we consider one of the possible methods of applying Lyapunov functions to stability problems. First we consider some simple auxiliary systems of equations for which it is possible to construct a Lyapunov function, or at least to establish the existence of a Lyapunov function v with definite properties. Then we investigate the maintenance of the particular properties (structural stability) for the more complicated equations of perturbed motion. On pp. 298–99 of his book [114], Lyapunov gave the first rigorous proof of stability in the neighborhood of an unperturbed motion in the case he called nonsingular. In the nonsingular case, Lyapunov succeeded in establishing the existence of a quadratic form that can serve as a testing function v [114, p. 276]. The mathematical statement of the description above is as follows. The perturbed system of equations is assumed to have the form

$$(18.1) \qquad \frac{dx_i}{dt} = X_i(x_1, \cdots, x_n, t) + R_i(x_1, \cdots, x_n, t) \qquad (i = 1, \cdots, n),$$

and it is known in some way that the trajectories of the auxiliary system

$$(18.2) \qquad \frac{dx_i}{dt} = X_i(x_1, \cdots, x_n, t) \qquad (i = 1, \cdots, n)$$

have a certain property (stability, instability, asymptotic stability). It is further supposed that a Lyapunov function v with these properties can be constructed for the system (18.2). We wish to know under what restrictions on the variation R_i in the right members of equations (18.1) of perturbed motion the function v will maintain the same properties.

This question can be stated rigorously in either of two (different) ways.

(a) Find a scalar function $R(x_1, \cdots, x_n)$ that is positive for $x \neq 0$, so that the character of the trajectories of (18.2) is maintained in the trajectories of (18.1) for all R_i such that $|R_i| \leq R$. That is, study a local question.

(b) Find realistic bounds on the initial functions R_i so that the trajectories of (18.1) have the same properties as the trajectories of (18.2), or at least find the order that the functions R_i must have. These are nonlocal questions, i.e., properties in the large.

The first of the problems mentioned above helps us to define a general type of property. When we can establish a positive answer to this problem, we can show that stability is maintained for all variations R_i that are suitably restricted. Systems that remain stable for all sufficiently small variations R_i are called *structurally stable*. It can be imagined that in proving that a system is structurally stable, there is considerable freedom in the choice of the function v in the set of all possible Lyapunov functions of the system (18.2), but the second problem is a more delicate one, and here a careful solution of the problem might require that the Lyapunov function of the system (18.2) be very carefully chosen.

In this chapter we consider certain characteristic problems of the two types mentioned. Although we shall be able to obtain fairly general results, possible application of our results to particular problems is restricted by two circumstances. (1) The main application is to those theorems of Lyapunov for which the derivative dv/dt is definite (and where, moreover, the corresponding properties are structural properties). (2) In estimating the admissible variations R_i, we use only the restriction $|R_i| \leq R$, and we do not consider the nature of the functions R_i in more detail than this.

19. Structural Stability and Instability

In investigating structural stability or instability we establish two facts:

(a) We prove that property A (definition 4.1, p. 19) is a structural property, and in particular, uniform asymptotic stability is a structural property, and instability is a structural property, if the instability is accompanied by property A.

(b) We show that in general instability is not a structural property.

It is fairly obvious that nonasymptotic stability cannot be a structural property. For consider the example $dx/dt = 0$. The null solution $x = 0$ is stable, but there are arbitrarily small functions $R(x)$ such that the null solution of the equation $dx/dt = R(x)$ is unstable whenever the function $R(x)$ satisfies $xR(x) > 0$, $x \neq 0$.

We now define the term *structural property*.

DEFINITION 19.1.   *Let a system of equations* (18.2) *be given. Suppose that tha trajectories of this system enjoy a certain property* P. *This property will be called a structural property in the neighborhood of the point* $x = 0$ (*in the region H*) *provided that in some neighborhood of the point* $x = 0$ *there exists a continuous function* $\eta(x_1, \cdots, x_n) > 0$, $x \neq 0$, *such that the perturbed system* (18.1) *also enjoys property* P *for all* R_i *that satisfy the inequality*

$$(19.1) \qquad\qquad |R_i(x, t)| \leq \eta(x) \qquad (i = 1, \cdots, n).$$

The contrary case is the following. If there is a system (18.2) that enjoys property P and is such that for every continuous function $\eta(x_1, \cdots, x_n)$ that is positive for $x \neq 0$, there exist functions R_i, depending on η and satisfying inequality (19.1), and for which, moreover, the corresponding system (18.1) does not enjoy property P, then property P is not a structural property.

Barbašin [13], [14] showed that asymptotic stability of the stationary equations of perturbed motion is a structural property if the functions $X_i(x)$ are continuously differentiable. If these functions are merely differentiable but periodic, asymptotic stability is a structural property [129], [130]. Asymptotic stability uniform with respect to the initial coordinates and time (x_0, t_0) was shown to be a structural property by Goršin [63], [64] and Malkin [125] for general functions X_i. It follows from results obtained by Kurcveil' [97] and Massera [131] that when the functions $X_i(x, t)$ are periodic and merely continuous, asymptotic stability is a structural property.

We begin with the following simple preliminary lemma.

LEMMA 19.1. *Let $\bar{h}_0$ be a closed region that does not necessarily contain the point $x = 0$. Let a function $v(x, t)$ have the following properties in $\bar{h}_0$:*

(i) *$v(x, t)$ has continuous partial derivatives $\partial v/\partial x_i$ $(i = 1, \cdots, n)$ that are uniformly bounded, $0 < t < \infty$.*

(ii) *The derivative $(dv/dt)_{(18.2)}$ evaluated along a solution of (18.2) is a function of definite sign.*

Then there is a function $\eta(x_1, \cdots, x_n)$ that is continuous, positive for $x \neq 0$ $[\eta(x) > 0]$, and such that the derivative $(dv/dt)_{(18.1)}$ evaluated along a solution of (18.1) is also a function of definite sign in the region $\bar{h}_0$, provided that $R_i(x, t)$ satisfies inequalities (19.1).

PROOF. Suppose for definiteness that the function $(dv/dt)_{(18.2)}$ is negative-definite. This means that there is a function $w(x)$ that is positive for $x \neq 0$ $[w(x) > 0]$, such that the relation

$$(19.2) \qquad \left(\frac{dv}{dt}\right)_{(18.2)} \leq -w(x)$$

holds for $x \in \bar{h}_0$, $t \in [0, \infty)$.

Write $N = \sup |\partial v/\partial x_i|$, $x \in \bar{h}_0$, $0 \leq t < \infty$; write $\eta(x) = (1 - q)w(x)/nN$. By using inequalities (19.1) and (19.2) we obtain

$$\left(\frac{dv}{dt}\right)_{(18.1)} = \sum_{i=1}^{n} \frac{\partial v}{\partial x_i}(X_i + R_i) + \frac{\partial v}{\partial t} = \left(\frac{dv}{dt}\right)_{(18.2)} + \sum_{i=1}^{n} \frac{\partial v}{\partial x_i} R_i$$

$$\leq -w(x) + \sum_{i=1}^{n} \left| \frac{\partial v}{\partial x_i} R_i \right| \leq -qw(x) \qquad (0 < q < 1)$$

for every x in the region $\bar{h}_0$, and the assertion of the lemma is established. The relation

$$(19.3) \qquad \eta(x) = \frac{(1 - q)w(x)}{nN} \qquad (0 < q < 1)$$

will be used below.

Now let us suppose that the functions X_i have the properties imposed in section 1. Through the use of lemma 19.1 in conjunction with theorems 4.2, 5.1, 6.1, and 6.2, the following theorem is easily established by methods already adumbrated. We omit the details.

THEOREM 19.1. *The property of the trajectories given in section 4 (p. 19) as property A, definition 4.1, is a structural property. If the null solution $x = 0$ is asymptotically stable or unstable, and property A subsists, then the stability or instability is a structural property.**

If the right members X_i of the equations of perturbed motion are periodic functions of the time t (all with the same period), or if they do not depend explicitly on the time, then the property of the system that no entire trajectory is contained in any neighborhood of the point $x = 0$ is a structural property. In this case, asymptotic stability is always a structural property.

It should be pointed out that the theorem (p. 33) on the existence of a Lyapunov function $v(x, t)$ in the case of asymptotic stability with respect to perturbations in a bounded region G_δ, theorem 5.3 on the existence of Lyapunov functions in the case of asymptotic stability for arbitrary initial perturbations, and the theorem on instability (6.1) permit us to prove also that for sufficiently small variations R_i, the region for which the trajectories have the corresponding property is preserved. In the case of asymptotic stability, a bounded region G_δ lying in the region of attraction of the point $x = 0$ remains entirely in this region; the property of uniform asymptotic stability for arbitrary initial perturbations is maintained; in the case of instability, the property of instability of the null solution $x = 0$ in the given region is preserved. Indeed, it follows from lemma 19.1 that all necessary facts for proof of these assertions on the properties of the Lyapunov functions of theorems 5.3 and 6.1 are preserved in the regions mentioned in those theorems. All these remarks on the persistence of the properties in the respective regions, moreover, refine the results cited above [13], [14], [63], [64], [97], [125], [129]–[131], where this question was not taken up in detail.

Now we give examples in which instability is *not* a structural property of the system, not only when the functions X are analytic, but even when it is known that the equilibrium point $x = 0$ is an isolated singular point of the system (18.2).

We consider the system of equations

$$\frac{dx}{dt} = y^3 + z^2 x ,$$

(19.4)
$$\frac{dy}{dt} = -x^3 + z^2 y ,$$

$$\frac{dz}{dt} = -z^3 .$$

The null solution $x = y = z = 0$ of this system is unstable. To demonstrate this, take for v the function $v = x^4 + y^4$, and compute as follows:

(19.5)
$$\frac{dv}{dt} = 4(x^4 + y^4)z^2 = 4z^2 v ,$$

---

* In the case of asymptotic stability, property A is equivalent to stability that is uniform with respect to t_0 and x_0.

along a solution of (19.4); thus, if we integrate (19.5) and use (19.4), we obtain

(19.6) $$v(x(t), y(t)) = v(x(0), y(0))z(0)^4 z^{-4}(t) .$$

From the third equation of (19.4) we see that if $z(t) \to 0$ as $t \to \infty$, then lim $v = \infty$ by (19.6). This is clearly sufficient to show that the null solution $x = y = z = 0$ of (19.4) is unstable.

Now let $\eta(x, y, z)$ be an arbitrary function subjected only to the conditions that it be continuous and that it be positive for $(x, y, z) \neq (0, 0, 0)$. Then we can always find functions $R_i(x, y, z)$ that satisfy the relations

$$| R_i(x, y, z) | \leq \eta(x, y, z), \qquad i = 1, 2, 3 ,$$

in some neighborhood of the point $x = y = z = 0$, and such that the perturbed system

(19.7) $$\frac{dx}{dt} = y^3 + z^2 x + R_1(x, y, z), \qquad \frac{dy}{dt} = -x^3 + z^2 y + R_2(x, y, z) ,$$

$$\frac{dz}{dt} = -z^3 + R_3(x, y, z)$$

is stable (in fact, is asymptotically stable).

Indeed, let ε be an arbitrary positive number. We construct the surface h consisting of that part of the surface

$$v = x^4 + y^4 = c$$

that lies interior to the sphere

(19.8) $$x^2 + y^2 + z^2 < \varepsilon^2 ,$$

where c is a sufficiently small positive number, $c < \varepsilon, c \ll \frac{1}{4}$. On this particular portion of the surface $v = c$, the function $\eta(x, y, z)$ has a positive minimum, $\phi(c)$ say. Let $f(x, y, z)$ be a continuous function (the existence of which is obvious), defined in a neighborhood of the point $x = y = z = 0$, having the property

(19.9) $$\frac{1}{2} \phi(c) \leq f(x, y, z) \leq \phi(c) \quad \text{on the surface } v = c ,$$

and satisfying the inequality $f(x, y, z) \leq \eta(x, y, z)$ in some neighborhood of the point $x = y = z = 0$.

Define the functions $R_i(x, y, z)$ by the relations

(19.10) $$R_1 = -fx, \quad R_2 = -fy, \quad R_3 = 0 .$$

For this choice of the functions R_i, the null solution $x = y = z = 0$ of the system (19.7) is clearly stable. To demontrate this, let $v = x^4 + y^4$. Along a trajectory of (19.7), we see by (19.10) that

(19.11) $$\left(\frac{dv}{dt} \right)_{(19.7)} = 4vz^2 - 4f(x^4 + y^4) ,$$

and that if $z^2 \leqq \phi(c)/4$, the inequality

$$(19.12) \qquad \left(\frac{dv}{dt}\right)_{(19.7)} \leqq -c\phi(c) < 0 \quad \text{for} \quad x^2 + y^2 + z^2 \neq 0$$

is valid on every surface $v = c$, by (19.9).

For $z \neq 0$, the third of inequalities (19.7) shows that $d(z^2)/dt < 0$. Thus the surface $v = c$, $z^2 = \phi(c)/4$ forms a closed shell that the trajectories of system (19.7) penetrate in a strictly inward direction. However, this means that this closed shell is a level surface of a Lyapunov function V, and that the upper right derivative

$$\lim_{\varDelta t \to +0} \sup \varDelta V / \varDelta t$$

of this function is negative-definite. Thus the null solution $x = y = z = 0$ of system (19.7), with R_i given by (19.10), is asymptotically stable. This makes it clear that the instability of this solution is not a structural property.

Let us continue the discussion of this example, and set $f(\mu, x, y, z) = \mu f(x, y, z)$, so that the function f depends continuously on the parameter μ. We see that it is possible to pass from a system in which the null solution is asymptotically stable to a system in which it is unstable, without passing through a condition of (ordinary) stability. It should be remarked that theorem 19.1 does show, however, that instability can fail to be a structural property only if property A does not hold. Thus, in the given (stationary) case, there must be an entire trajectory ($\neq 0$) of the system in every neighborhood of the point $x = y = z = 0$. Such a trajectory for (19.4) is the closed curve $x^4 + y^4 = \text{const.}$, $z = 0$.

If the functions X_i do not depend on the time explicitly, theorems 12.1 and 12.2 make it possible to give a sufficient condition that instability be of the structural variety. We suppose that the right members $X_i(x)$ of the system (19.13) have continuous partial derivatives $\partial X_i/\partial x_j$ in some neighborhood of the point $x = 0$.

THEOREM 19.2. *Suppose that the system of equations*

$$(19.13) \qquad \frac{dx_i}{dt} = X_i(x_1, \cdots, x_n), \qquad i = 1, \cdots, n,$$

has a solution $x(x_0, t)$ which comes arbitrarily close to the point $x = 0$ for $t \to -\infty$. Then instability of the null solution $x = 0$ of the system (19.13) will be structural. Moreover, the very property that the system have a trajectory which converges on the point $x = 0$ for $t \to -\infty$ is a structural property.

The details of the proof are omitted; the theorem follows in an obvious way from theorems 12.1 and 12.2 by use of lemma 19.1.

This concludes our study of the general question of structural properties of stability and instability. In the subsequent sections of this chapter, we consider structural systems for which the admissible variations $R_i(x, t)$ satisfy order estimates.

20. Criteria for Asymptotic Stability of Quasi-linear Systems

In this section we shall assume that the functions X_i which appear in the right members of the equations of motion have continuous partial derivatives $\partial X_i/\partial x_j$, and that the latter satisfy the inequalities $|\partial X_i/\partial x_j| \leq L$ for all x, $\|x\| < \infty$.

Let us suppose that the solutions $x(x_0, t_0, t)$ of the approximating system of equations (18.2) satisfy the inequality

$$(20.1) \qquad \|x(x_0, t_0, t)\|_2 \leq B\|x_0\|_2 \exp[-\alpha(t - t_0)] \quad \text{for} \quad t \geq t_0$$

for all initial conditions x_0, t_0. By theorem 11.1, condition (20.1) is equivalent to the hypothesis that a function $v(x, t)$ exists that satisfies the following relations [along (18.2)]:

$$
\begin{aligned}
&c_1\|x\|_2^2 \leq v(x, t) \leq c_2\|x\|_2^2, \\
(20.2) \qquad &\left(\frac{dv}{dt}\right)_{(18.2)} \leq -c_3\|x\|_2^2, \\
&\left|\frac{\partial v}{\partial x_i}\right| \leq c_4\|x\|_2,
\end{aligned}
$$

where c_1, c_2, c_3, c_4 are positive constants. By lemma 19.1, using estimate (19.3), we conclude that the properties of the function $v(x, t)$ are valid also along the trajectories of the system (18.1), provided that the functions $R_i(x, t)$ satisfy the inequalities

$$(20.3) \qquad |R_i(x, t)| < \frac{(1 - q)c_3\|x\|_2}{nc_4} \qquad (i = 1, \cdots, n).$$

The form of the second inequality of (20.2) is unaltered, but its exact statement becomes

$$(20.4) \qquad \left(\frac{dv}{dt}\right)_{(18.1)} \leq -qc_3\|x\|_2^2.$$

By using the fundamental results of section 11, and applying lemma 19.1 in the proper manner, we arrive at the following assertion.

THEOREM 20.1. *If the solutions of system* (18.2) *satisfy conditions* (20.1), *or equivalently if a function* $v(x, t)$ *exists that satisfies the estimates* (20.2), *then a corresponding set of estimates will hold for the solutions of the system* (18.1) *(but with altered constants), provided the functions* R_i *satisfy the inequality* (20.3).

Remark. Suppose that the inequality (20.3) or the estimates (20.2) are satisfied only in the region

$$(20.5) \qquad \|x\|_2 \leq R.$$

Then relation (20.1) will be satisfied by solutions of the system (18.1) for arbitrary initial values t_0, $0 < t_0 < \infty$, if x_0 lies in the region

$$(20.6) \qquad v(x_0, t_0) < \inf[v(x, t) \quad \text{for} \quad \|x\|_2 = R] = c_1 R^2$$

and *a fortiori* if x_0 lies in the region

(20.7) $$\| x_0 \|_2^2 < \frac{c_1}{c_2} R^2 .$$

To demonstrate this, note that the trajectories $x(x_0, t_0, t)$ of the system (18.1) cannot leave the region (20.5) if conditions (20.3) are satisfied and if x_0 lies in the region (20.7). This is true because the lemma shows that estimate (20.4) holds in the region (20.5), and because the values of $v(x, t)$ on the boundary of the same region exceed $v(x_0, t_0)$.

We can then integrate inequality (20.4) by the use of the first line of (20.2), and obtain relation (20.1). Thus the assertion is established; of course, the constants in relation (20.1) may be different for equations (18.1) than for equations (18.2).

This theorem includes as a special case the criteria for first-order asymptotic stability of Lyapunov [114, p. 297], and of K. P. Persidskii [145]. The statement of the theorem shows that asymptotic stability of a solution of a quasi-linear (close-to-linear) system is not dependent on the fact that the approximating system (18.2) is linear, but only on the fact that the solutions of the approximating system satisfy conditions (20.1).

In [18], Barbašin and Skalkina used another method in studying the importance of relations (20.1) for the stability of solutions of nonlinear systems. The author [88] used Lyapunov functions to study the same problem.

We obtained estimate (19.3) under very general assumptions. To solve concrete problems, we shall try to obtain better estimates, which can in fact be obtained if we can make a successful choice of the function $v(x)$.

We shall be able to use the very simple equation

(20.8) $$\frac{dx_i}{dt} = d_{i1}x_1 + \cdots + d_{in}x_n \qquad (i = 1, \cdots, n)$$

as the auxiliary system (18.2). This is especially advantageous, since a Lyapunov function for this system is a quadratic form in the arguments $x_1, \cdots, x_n$. (For the system of equations

(20.9) $$\frac{dx_i}{dt} = Y_i(x_1, \cdots, x_n, t) \qquad (i = 1, \cdots, n)$$

to admit system (20.8) as an approximating system, it is obviously necessary that the system (20.9) be almost linear in some sense.) This approach is one of the most-used methods for studying nonlinear problems [1], [2], [4], [5], [9], [66], [122], [123], [124], [168], [175]–[177], [179], [195].

We give three examples for constructing Lyapunov functions $v(x, t)$. They are to be found in the work of Četaev [36].

Suppose a system of equations (20.9) has the form

(20.10) $$\frac{dx_i}{dt} = p_{i1}(t)x_1 + \cdots + p_{in}(t)x_n + \varphi_i(x_1, \cdots, x_n, t) \qquad (i = 1, \cdots, n) ,$$

where $p_{ij}(t)$ are continuous or piecewise-continuous bounded functions of the time.

Case 1. Suppose the limit

(20.11) $\lim_{t \to \infty} p_{ij}(t) = d_{ij}$

exists. Suppose further that the roots of the characteristic equation

(20.12) $\det [d_{ij} - \lambda \delta_{ij}]_1^n = 0$

have negative real part. Then system (20.8) is a suitable auxiliary system.

Case 2. Suppose the functions $p_{ij}(t)$ are periodic functions of the time t, all with the same period ϑ. The auxiliary system (20.8) is that for which the coefficients d_{ij} are defined by the formula

(20.13) $d_{ij} = \frac{1}{\vartheta} \int_0^\vartheta p_{ij}(\xi) \, d\xi$.

Case 3. Let $t = t_0 > 0$ be some moment of time that is physically important in studying the system (20.10). The auxiliary system (20.8) is defined by the relation

(20.14) $d_{ij} = p_{ij}(t_0)$.

In cases 1, 2, and 3 it is supposed that the roots λ_i of equation (20.12) satisfy an inequality

(20.15) $\mathrm{Re}\, \lambda_i < -\delta < 0$,

where δ is a fixed positive number. After defining the auxiliary system (20.8) we proceed as follows. We take a negative-definite form

(20.16) $w(x) = \sum_{i,j=1}^n c_{ij} x_i x_j \qquad (c_{ij} = c_{ji})$

and construct a function $v(x) = \sum a_{ij} x_i x_j$ that satisfies the relations

(20.17) $\left(\dfrac{dv}{dt} \right)_{(20.8)} = w(x)$.

(See p. 58.) To compute the value of $(dv/dt)_{(20.10)}$ along trajectories of the complete system (20.10), we find

(20.18) $\left(\dfrac{dv}{dt} \right)_{(20.10)} = w(x) + \sum_{i,j,k=1}^n [a_{ij}(p_{jk} - d_{jk})$

$+ a_{kj}(p_{ji} - d_{ji})] x_i x_k + \sum_{i,j=1}^n a_{ij} x_j \varphi_i(x, t)$.

It is clear that we can assume the functions φ_i have the form[*]

(20.19) $\varphi_i(x_1, \cdots, x_n, t) = \sum_{k=1}^n h_{ik}(x, t) x_k$,

since the derivative dv/dt evaluated along a trajectory of (20.10) has the value

---

[*] A linear transformation of the arguments from x_i to $y_i = \sum b_{ij} x_j$ will make it possible to simplify the analysis considerably when the functions φ_i are written in the form (20.19). See Rumyancev [168].

(20.20) $$\left(\frac{dv}{dt}\right)_{(20.10)} = \sum_{i,j,k=1}^{n} [c_{ik} + a_{ij}(p_{jk} - d_{jk}) + a_{kj}(p_{ji} - d_{ji})$$
$$+ a_{ij}h_{jk} + a_{kj}h_{ji}]x_i x_k .$$

If the form in the right member of (20.20) is negative-definite as a function of the arguments x_i, the null solution $x = 0$ of (20.10) will be asymptotically stable.

A general theorem of Sylvester proved in [60a, p. 88] states that a quadratic form $\sum e_{ik}x_i x_k = -dv/dt$ is positive-definite if the nested set of principal minors (the leading principal minors) of the matrix (e_{ik}) are all positive:

(20.21) $$\Delta_n > 0, \quad \Delta_{n-1} > 0, \quad \cdots, \quad \Delta_1 > 0 .$$

In the special case under consideration, the coefficients e_{ik} are variable. Thus (20.21) is not a sufficient condition for negative-definiteness of dv/dt. For this, the stronger condition

(20.22) $$\Delta_n > \gamma, \quad \cdots, \quad \Delta_1 > \gamma \quad (\gamma > 0)$$

would be needed. Here, γ is a positive constant independent of x, t. The system of inequalities (20.22) determines bounds for the parameters h_{ij}, $(p_{ij} - d_{ij})$ for which the null solution $x = 0$ of (20.10) is stable.

One way to obtain lower bounds for values of $|h_{ij}|, |p_{ij} - d_{ij}|$ for which the null solution $x = 0$ of (20.10) is stable is the following. Set

(20.23) $$p_{ij} - d_{ij} = \mu q_{ij}, \quad h_{ij} = \mu g_{ij} .$$

Then (20.20) takes the form

(20.24) $$\left(\frac{dv}{dt}\right)_{(20.10)} = \sum_{i,j,k=1}^{n} (c_{ij} + \mu[a_{ij}(q_{jk} + g_{jk}) + a_{kj}(q_{ji} + g_{ji})])x_i x_k .$$

When $\mu = 0$, the quadratic form (20.24) is negative-definite by the choice of the c_{ij}. Thus if $\mu = \mu_0$ is a value of μ for which the function (20.24) has a negative sign, this value of μ when used in (20.23) gives a lower bound for the values of $p_{ij} - d_{ij}, h_{ij}$ for which stability of the system (20.10) is assured. Now if μ varies continuously, the coefficients of the quadratic form (20.24) vary continuously. If the inequalities (20.21) are all satisfied for some value of μ, then as μ decreases, the first one to be violated is the inequality $\Delta_n > 0$. Thus the bound μ_0 can be taken to be the root of the determinantal equation

$$\det \begin{bmatrix} c_{11} + 2\mu \sum_{j=1}^{n} [a_{1j}(q_{j1} + g_{j1})] & \cdots & c_{1n} + \mu \sum_{j=1}^{n} [a_{1j}(q_{jn} + g_{jn}) \\ & & + a_{nj}(q_{j1} + g_{j1})] \\ \cdots\cdots\cdots\cdots\cdots\cdots\cdots\cdots\cdots\cdots\cdots\cdots\cdots\cdots\cdots & & \\ c_{n1} + \mu \sum_{j=1}^{n} [a_{nj}(q_{j1} + g_{j1}) \\ \quad + a_{1j}(q_{jn} + g_{jn})] & \cdots & c_{nn} + 2\mu \sum_{j=1}^{n} [a_{nj}(q_{jn} + g_{jn})] \end{bmatrix} = 0$$

that has smallest modulus.

If the null solution is to be *asymptotically* stable, the derivative dv/dt (20.24) must be *positive-definite*. If the above discussion is slightly changed by using an additional parameter ε to write

$$\left(\frac{dv}{dt}\right)_{(20.10)} = \sum_{i,j,k=1}^{n} (c_{ik} + \varepsilon\delta_{ik} + \mu[a_{ij}(q_{jk} + g_{jk}) + a_{kj}(q_{ji} + g_{ji})])x_i x_k - \varepsilon \sum_{i=1}^{n} x_i^2$$

$$(\varepsilon > 0),$$

it will be seen that a sufficient condition for asymptotic stability of the null solution $x = 0$ of the system (20.10) is that $|\mu| < |\mu_0|$, where μ_0 is that root of the determinantal equation

$$\det \begin{bmatrix} c_{11} + \varepsilon + 2\mu\sum_{j=1}^{n} a_{1j}(q_{j1} + g_{j1})] & \cdots & c_{1n} + \mu\sum_{j=1}^{n}[a_{1j}(q_{jn} + g_{jn}) \\ & & + a_{nj}(q_{j1} + g_{j1})] \\ \cdots\cdots\cdots\cdots\cdots\cdots\cdots\cdots\cdots\cdots\cdots\cdots\cdots\cdots \\ c_{n1} + \mu\sum_{j=1}^{n}[a_{nj}(q_{j1} + g_{j1}) \\ \quad + a_{1j}(q_{jn} + g_{jn})] & \cdots & c_{nn} + \varepsilon + 2\mu\sum_{j=1}^{n}[a_{nj}(q_{jn} + g_{jn})] \end{bmatrix} = 0$$

which has smallest modulus. Moreover, if R_0, R are related by the condition

(20.25) $\sup[v(x)$ for $||x|| < R_0] < \inf[v(x)$ for $||x||_2 = R]$,

then the region $||x||_2 < R_0$ will lie in the region of attraction of the point $x = 0$, provided the inequality (20.22) is satisfied in the region $||x||_2 \leq R$.

If the number R is given, a bound for the number R_0 can be obtained as follows. If $\rho_{\min}$ and $\rho_{\max}$ are respectively the moduli of the roots of smallest and greatest modulus of the equation $\det[a_{ij} - \delta_{ij}\rho] = 0$, the function $v(x)$ satisfies the inequalities

(20.26) $\rho_{\min}||x||_2^2 \leq v(x) \leq \rho_{\max}||x||_2^2$.

Thus the estimate $R_0 = R$ is valid, because of (20.25) and (20.26).

The sufficient conditions (20.22) for asymptotic stability of the null solution $x = 0$ depend on the choice of the function $v(x)$, or (what is the same thing) the choice of the function $w(x)$ used to construct the auxiliary system (20.8). By different choices of these functions v, w, sharper bounds for asymptotic stability may be obtained, and in the same way, less restrictive radii R_0 for the region of attraction. Staržinskiĭ [175]–[177] studied this question in detail for certain second-, third-, and fourth-order linear systems with variable coefficients.

21. Sufficient Conditions for the Asymptotic Stability of Solutions of Nonlinear Systems of Differential Equations

In this section, we give criteria for the asymptotic stability of the null solution $x = 0$ of a nonlinear system

$$(21.1) \qquad \frac{dx_i}{dt} = X_i(x_1, \cdots, x_n, t) \qquad (i = 1, \cdots, n) \,,$$

where the functions X_i are defined and continuous in all space and for all values of t, and where they have continuous partial derivatives $\partial X_i/\partial x_j$ that are uniformly bounded in every bounded region:

$$(21.2) \qquad \left| \frac{\partial X_i}{\partial x_j} \right| < L_R \qquad (i = 1, \cdots, n; \ j = 1, \cdots, n) \,,$$

whenever $\| x \|_2 < R$.

In section 20, we discussed the stability of solutions of (20.22) on the basis of boundedness of the functions X_i. In this section, the criteria we use have to do with the boundedness of the derivatives of the X_i. These criteria are based on theorem 16.2, when we take the function $v(x_0, t_0, z, t)$ of that theorem to be a quadratic form

$$(21.3) \qquad v(x_0, t_0, z, t) = \sum_{i,j=1}^{n} a_{ij} z_i z_j \,.$$

In fact, we shall investigate the asymptotic stability of the perturbed trajectories $x(x_0, t_0, t)$ by methods used in the preceding section. The criteria we develop were given in [85], [94], and (in a special case) [195].

THEOREM 21.1. *Let* (a_{ij}) *be an* $n \times n$ *matrix of constants with the following properties*:

(i) (a_{ij}) *is symmetric*;

(ii)   *the proper values* $\rho_1, \cdots, \rho_n$ *of this matrix are positive*;

(iii)   *the proper values of the symmetric matrix* (c_{ik}) *given by*

$$(21.4) \qquad c_{ik} = \sum_{j=1}^{n} \left(a_{ij} \frac{\partial X_j}{\partial x_k} + a_{kj} \frac{\partial X_j}{\partial x_i} \right)$$

are uniformly negative; i.e., these proper values $\lambda_1, \cdots, \lambda_n$ *uniformly satisfy the conditions*

$$(21.5) \qquad \lambda_i < -\gamma \quad (\gamma > 0; \ i = 1, \cdots, n)$$

for all x *and all* t, *where* γ *is constant.*

Then the null solution $x = 0$ *of equation* (21.1) *is asymptotically stable for arbitrary initial perturbations.*

In the literature, this type of stability has sometimes been called *global*. This usage is adopted in this book. The term *stability in the large* is reserved for asymptotic stability that enjoys a certain kind of uniformity; see p. 30.

Remark. If the system (21.1) is linear, i.e., has the form (20.8), the conditions of the theorem are not only sufficient but also necessary. This assertion is proved as follows. If these conditions are satisfied, the function $v(x) = \sum a_{ij} x_i x_j$ is a Lyapunov function for the system, and its derivative

(21.6)
$$\frac{dv}{dt} = \sum_{i,k=1}^{n} c_{ik} x_i x_k = \sum_{i,j,k=1}^{n} (a_{ij} d_{jk} + a_{kj} d_{ji}) x_i x_k$$

is negative-definite, owing to condition (21.5). The general criterion of Lyapunov for asymptotic stability is thus satisfied, and the null solution $x = 0$ is asymptotically stable. On the other hand, if this solution of (20.8) is asymptotically stable, then by Lyapunov's theorem [114, p. 276] there is a function $v(x) = \sum a_{ij} x_i x_j$ that is positive-definite; moreover, the derivative dv/dt of this function is negative-definite along a solution of (20.8), and all the conditions of theorem 21.1 are satisfied.

We note further that the matrix (a_{ij}) can be chosen to be the unit matrix $a_{ii} = 1$, $a_{ij} = 0$ $(i \neq j)$. In this case, the matrix (c_{ik}) will be the matrix defined by

(21.7)
$$c_{ik} = \frac{\partial X_i}{\partial x_k} + \frac{\partial X_k}{\partial x_i} .$$

Thus the theorem asserts that the null solution is asymptotically stable for arbitrary initial perturbations x_0 whenever the matrix (21.7) satisfies conditions (21.5). The theorem of Markus and Yamabe [128] includes this theorem as a corollary. Indeed, Markus and Yamabe allow γ to be a function of $||x||$, $\gamma = \nu(||x||)$, and derive the conclusion of the present theorem by imposing on the function ν the condition that $\nu(\sigma)$ is monotonically decreasing, and that for each $\varepsilon > 0$, the relation

$$\int_0^\infty \exp\left\{-\varepsilon \int_0^\rho \nu(\sigma) d\sigma\right\} d\rho < \infty$$

holds.

PROOF OF THE THEOREM. The method of proof is to establish the fact that the function $v(x_0, t_0, z, t)$ defined by (21.3) satisfies the conditions of theorem 16.2.

We deal with a trajectory $x(x_0, t_0, t)$ of system (21.3) as follows. From it we construct an equation of perturbed motion, assuming that the given trajectory is an unperturbed motion. Thus the new dependent variable must be taken as $z(t) = x(t) - x(x_0, t_0, t)$:

(21.8)
$$\frac{dz_i}{dt} = Z_i(z_1, \cdots, z_n, t)$$
$$= X_i(x_1, \cdots, x_n, t) - X_i(x_1(x_0, t_0, t), \cdots, x_n(x_0, t_0, t), t)$$
$$= \sum_{j=1}^{n} \left(\frac{\partial X_i}{\partial x_j}\right) z_j + R_i(z_1, \cdots, z_n, t) \qquad (i = 1, \cdots, n).$$

Here, the partial derivatives $\partial X_i/\partial x_j$ must all be evaluated at the points of the trajectory $x(x_0, t_0, t)$. Let $\beta > 0$ be an arbitrary positive number. The functions $\partial X_i/\partial x_j$ are continuous and hence uniformly continuous in every closed region $||x|| \leq H$, $t_0 \leq t \leq t_0 + T$. Thus there will exist a positive number $\delta(\beta, t_0, T, H)$ such that the inequality

(21.9)
$$|R_i(z, t)| < \beta ||z||_2$$

will be satisfied in the region

(21.10) $||z||_2 < \delta(\beta, t_0, T, H)$

as long as the point $x(x_0, t_0, t)$ satisfies $||x(x_0, t_0, t)|| \leq H$.

The value of the derivative dv/dt of the function $v(x_0, t_0, z(t), t)$ along a trajectory of the system (21.8) is given by

$$\frac{dv}{dt} = \sum_{j=1}^{n} \sum_{i,k=1}^{n} \left(a_{ij} \frac{\partial X_j}{\partial x_k} + a_{kj} \frac{\partial X_j}{\partial x_i} \right) z_i z_k + \sum_{j=1}^{n} \frac{\partial v}{\partial z_j} R_j .$$

Thus, by the use of inequalities (21.5) and (21.9), we see that the estimate

$$\frac{dv}{dt} \leq -\gamma ||z||_2^2 + \sum_{j=1}^{n} \left| \frac{\partial v}{\partial z_j} R_j \right| \leq -\gamma ||z||_2^2 + n\beta ||z||_2 \sup \left| \frac{\partial v}{\partial z_i} \right|$$

is valid in the region (21.10).

But the derivative $\partial v/\partial z_i$ clearly satisfies the inequality

$$\left| \frac{\partial v}{\partial z_i} \right| \leq n ||z||_2 \max |a_{ij}| = Nn ||z||_2 ,$$

and therefore we can draw the conclusion

(21.11) $\dfrac{dv}{dt} \leq [-\gamma + n^2 N\beta] ||x||_2^2$ $(N = \max |a_{ij}|)$.

If the number β is chosen to satisfy the relation $\beta = \gamma/(2n^2 N)$, the conclusion is that the relation

(21.12) $\dfrac{dv}{dt} \leq -\dfrac{\gamma}{2} ||z||_2^2$

is valid everywhere in the region (21.10).

Thus, the function v (21.3) does satisfy all the conditions of theorem 16.2, which is therefore proved.

Remark. The details of the proofs of theorems 16.2 and 21.1 show that asymptotic stability of the null solution $x = 0$ of the system (21.1) with respect to initial perturbations x_0, $||x_0|| < H_0$ follows if inequality (21.5) is satisfied in the region $||x_0|| < H_0$, where $H = H_0 \sqrt{(\rho_{max}/\rho_{min})}$. Here ρ_{max}, ρ_{min} are the greatest and least proper values of the matrix (a_{ij}).

Staržinskiĭ [175]–[177] investigated linear systems

(21.13) $\dfrac{dx_i}{dt} = \sum_{j=1}^{n} p_{ij}(t) x_j$

of second, third, and fourth order. In particular, he found conditions on the coefficients $p_{ij}(t)$ that would be sufficient to assure the existence of a Lyapunov function that is a quadratic form [satisfying the further conditions (20.2)]. His results are clearly contained in theorem 21.1, since the question amounts to the existence of a positive-definite matrix (a_{ij}) with the property that the matrix (c_{ij}), $c_{ik} = a_{ij} p_{jk} + a_{kj} p_{ji}$ has negative proper values. In fact, our general result uses $\partial X_i/\partial x_j$ in the place of $p_{ij}(t)$, and is equal to the latter in the cases considered by Staržinskiĭ [176], [177].

22. Theorems on mth-order Stability

In this section, we consider the system

$$(22.1) \quad \frac{dx_i}{dt} = X_i^{(m)}(x_1, \cdots, x_n, t) + R_i(x_1, \cdots, x_n, t) \qquad (i = 1, \cdots, n;\ m \geq 1),$$

where the functions $X_i^{(m)}(x, t)$ are homogeneous functions of the arguments $x_1, \cdots, x_n$ of degree m (for every positive value of $t \geq 0$), and satisfy the Lipschitz conditions

$$| X_i^{(m)}(x'', t) - X_i^{(m)}(x', t)| \leq L \| x'' - x' \| \quad \text{for} \quad \| x' \| \leq 1,\ \| x'' \| \leq 1 .$$

Since the X_i are assumed to be homogeneous, these conditions amount to the equivalent conditions

$$(22.2) \quad | X_i^{(m)}(x'', t) - X_i^{(m)}(x', t)| \leq Lr^{m-1} \| x'' - x' \| \quad \text{for} \quad \| x'' \| < r,\ \| x' \| < r .$$

We assume that in some neighborhood

$$(22.3) \qquad\qquad \| x \|_2 < H,\quad t > 0$$

the functions $R_i(x, t)$ satisfy the inequalities

$$(22.4) \qquad\qquad R_i(x, t) \leq \gamma \| x \|_2^m \qquad (i = 1, \cdots, n) .$$

When the number $\gamma > 0$ is sufficiently small, this condition is the usual one that the order of the functions R_i exceeds m in the region (22.3), i.e., that the order of the perturbing term in the right member of (22.1) exceeds the order of the main term $X_i^{(m)}$.

The practical value of this section comes from the fact that we relate (22.1) to the system

$$(22.5) \qquad\qquad dx_i/dt = X_i^{(m)}(x_1, \cdots, x_n, t) \qquad (i = 1, \cdots, n) ,$$

which is of course nonlinear in general. The important questions concern the conditions under which asymptotic stability (or instability) of the null solution $x = 0$ of equations (22.5) guarantees the stability (or instability) of the related system (22.1). If the conditions of property A (definition 4.1) are satisfied in the region (22.3), the results of section 19 show that the system (22.5) is close to the system (22.1) in the following sense. The null solution is asymptotically stable, or is unstable for (22.1), provided the corresponding property is enjoyed by (22.5), and provided further that the functions $|R_i|$ admit certain positive bounds, namely those of section 19. The point of the present section is to refine the bounds in question, i.e., to determine γ in relation (22.4) so that the results mentioned above continue to hold. The point of attack will be to obtain precise estimates for the function $v(x, t)$ that we know exists when system (22.5) has property A.*

---

* If the system (22.5) does not have property A, the instability of the null solution $x = 0$ is not a structural property. See example (19.7), p. 84.

We first consider the case in which the functions X_i do not depend explicitly on the time t. We begin by stating without proof a precise form of theorem 4.2 that holds when the right members X_i are homogeneous functions of degree $m \geq 1$. The proof of the theorem is found in the author's paper [88].

THEOREM 22.1. *Suppose that no solution* $(\not\equiv 0)$ *of system* (22.5) *is bounded,* $-\infty < t < \infty$. *Then a continuous differentiable function* $v(x_1, \cdots, x_n)$ *exists that satisfies the following inequalities:*

(22.6)
$$| v(x) | \leq c_2 || x ||_2^A ,$$
$$\left(\frac{dv}{dt} \right)_{(22.5)} \leq -c_3 || x ||_2^{A+m-1} ,$$
$$\left| \frac{\partial v}{\partial x_i} \right| \leq c_4 || x ||_2^{A-1} ,$$

where A, c_2, c_3, c_4 *are positive constants. If the null solution* $x = 0$ *is unstable, every neighborhood of the point* $x = 0$ *contains a point* $x_0 \neq 0$ *at which* $v(x_0) < 0$. *If the null solution* $x = 0$ *is asymptotically stable, the function* $v(x)$ *satisfies the further inequality*

(22.7)
$$c_1 || x ||_2^A \leq v(x) ,$$

where c_1 *is a positive constant.*

The following assertion is an easy corollary of theorem 22.1.

COROLLARY 22.1. *Let* $x = 0$ *be the only solution of system* (22.5) *that is bounded,* $-\infty < t < \infty$. *Then there is a positive constant* $\gamma > 0$ *such that whenever inequality* (22.4) *holds, the trajectories of* (22.1) *and* (22.5) *have the same character; that is, if the trajectories of one are asymptotically stable, so are the trajectories of the other; if the trajectories of one are unstable, so are the trajectories of the other.*

PROOF. Set $\gamma = (1 - q)c_3/(nc_4)$. The function $v(x)$ of theorem 22.1 will satisfy estimate (22.6) along a solution of system (22.1), by lemma 19.1. [See also (19.3).] Indeed, by using the second line of estimate (22.6), we find

(22.8)
$$\left(\frac{dv}{dt} \right)_{(22.1)} \leq -qc_3 || x ||_2^{A+m-1} \qquad (0 < q < 1) .$$

If the null solution $x = 0$ of system (22.5) is unstable, theorem 22.1 shows that every neighborhood of the point $x = 0$ contains a point $x_0 \neq 0$ for which $v(x_0) < 0$. Thus the function v satisfies all the conditions of Lyapunov's first theorem on instability for system (22.1); the null solution $x = 0$ is unstable.

On the other hand, if the null solution $x = 0$ of system (22.5) is asymptotically stable, the function v is positive-definite. Thus the function v is a Lyapunov function for system (22.1), and the null solution $x = 0$ of (22.1) is asymptotically stable. The proof of the corollary is complete.

Remark. Suppose the right members of the equations of perturbed motion are holomorphic functions of the arguments $x_1, \cdots, x_n$, expansible as a convergent series of such functions of degrees $m, m + 1, \cdots$. Corollary 22.1 gives a way of determining the asymptotic stability or instability of the null solution by considering only the lowest-order terms (of degree m).

The first proof of such a theorem was given by Malkin [120], [121], who confined himself to the case in which the functions $X_i^{(m)}(x)$ are forms of degree $m \geq 1$, and the functions $R_i(x, t)$ satisfy conditions (22.4).

It is important to note that if $m > 1$, and if the null solution $x = 0$ of system (22.5) is asymptotically stable, any solution $x(x_0, t)$ of this system will satisfy an inequality

$$(22.9) \qquad || x(x_0, t) ||_2^{m-1} \leq [B || x_0 ||_2^{1-m} + \alpha(t - t_0)]^{-1}$$

for $t \geq t_0$, where t_0, α, B are positive constants. The proper formulation for the case $m = 1$ is inequality (11.9), p. 59.

Inequality (22.9) is established as follows. Starting with (22.6) and (22.7), we obtain $dv/dt \leq -c_3 [v/c_2]^{(A+m-1)/A}$; integrating this, we obtain

$$-v\big(x(x_0, t)\big)^{(1-m)/A} + v(x_0)^{(1-m)/A} \leq -\beta(t - t_0) , \qquad \beta \text{ const.}, \ \beta > 0 .$$

From (22.6) and (22.7) we also obtain

$$c_1^{(1-m)/A} || x(x_0, t) ||_2^{1-m} \geq c_2^{1/A} || x_0 ||_2^{1-m} + \beta(t - t_0) ,$$

which, when written in the form

$$|| x(x_0, t) ||_2^{m-1} \leq [(c_2/c_1)^{(1-m)/A} || x_0 ||^{1-m} + \beta c_1^{(m-1)/A}(t - t_0)]^{-1} ,$$

establishes assertion (22.9).

Let us consider asymptotic stability of the null solution $x = 0$ of equation (22.1) in the general case. (Considerations needed for instability are similar.) Now if $m = 1$ (i.e., for linear equations), the asymptotic stability of the system (22.5) can be destroyed if functions R_i are added that are of arbitrarily small order, as explained on pp. 59 and 86. However, for linear equations with variable coefficients, if the order of the functions R_i is smaller than the order of the norm $|| x ||_2$ and if as the time t increases, the perturbed trajectories approach the unperturbed trajectory $x = 0$ exponentially [i.e., if inequality (11.9) is valid], then a linear system with variable coefficients that is asymptotically stable will remain asymptotically stable when perturbed by the addition of the functions R_i. See sections 11 (p. 59) and 20 (page 86).

When $m > 1$, the corresponding question concerns the asymptotic stability of system (22.5) and of system (22.1). The characterization of stationarity is that the trajectories $x(x_0, t_0, t)$ of the system (22.5) satisfy the inequality

$$(22.10) \qquad || x(x_0, t_0, t) ||_2^{m-1} \leq [B || x_0 ||_2^{1-m} + \alpha(t - t_0)]^{-1} .$$

It is also assumed that the functions R_i satisfy inequality (22.4) when the constant $\gamma > 0$ is sufficiently small.

The following theorem gives a positive answer to this question.

THEOREM 22.2. *Let the null solution $x = 0$ of the system* (22.5) *satisfy inequality* (22.10). *For some positive number $\gamma > 0$, the null solution $x = 0$ of system* (22.1) *will also be asymptotically stable for all R_i for which relation* (22.4) *holds.*

PROOF. First note that condition (22.10) shows that there is a function $v(x, t)$ that satisfies estimates (22.6) and is positive-definite; indeed this function $v(x, t)$ satisfies inequality (22.7). Proof of this assertion in the general case involves a mountain of detail. In the special case in which the functions $X_i^{(m)}(x, t)$ possess continuous partial derivatives $\partial X_i/\partial x_j$, the proof was given in [88]. By complicating this same proof, it is possible to demonstrate the existence of a continuously differentiable function $v(x, t)$ that satisfies estimates (22.6) and (22.7) on the basis of the weaker assumption that the functions $X_i^{(m)}(x, t)$ satisfy only a Lipschitz condition (22.2).

The proof of the theorem is completed as follows. Lemma 19.1 and estimate (19.3) show that, since the function $v(x, t)$ satisfies estimates (22.6) and (22.7), it must have a derivative dv/dt that is negative-definite along a trajectory of (22.1), provided that the functions R_i satisfy the inequality (22.4) for sufficiently small $\gamma > 0$. Thus the same function $v(x, t)$ is a Lyapunov function for the entire system (22.1), and the null solution $x = 0$ of this system must be asymptotically stable.

5

Applications of Lyapunov's Method to the Solution of Some Special Problems in Stability

23. Preliminary Remarks

The method expounded in Chapter 4 for constructing a Lyapunov function for practical systems is the simplest and most direct method. Nevertheless, this method has some drawbacks. The derivative dv/dt will enjoy the required properties for the complete system of equations only when the moduli of the variations R_i are bounded by very narrow bounds. The approach of Chapter 4 made a result of this kind inevitable, since no notice was taken of the particular form of the variations R_i. Nevertheless, in certain cases, we have additional information about the variations R_i and can use this information to construct a Lyapunov function that is suitable for the case at hand. In this way, we can sometimes obtain sufficient criteria for stability that are rather broad, perhaps as broad as is practically desirable. The same remarks apply also to conditions for instability. In constructing a Lyapunov function for problems in which the variations R_i are restricted in some manner other than in modulus, the Lyapunov function must depend in some way on the functions R_i.

Let us consider a system of equations of perturbed motion

$$(23.1) \qquad \frac{dx_i}{dt} = X_i(x_1, \cdots, x_n, t, \quad \beta_1, \cdots, \beta_k,$$

$$\varphi_1(x_1, \cdots, x_n, t), \cdots, \varphi_r(x_1, \cdots, x_n, t)),$$

the right member of which contains known functions of the arguments $x_1, \cdots, x_n, t$, parameters $\beta_1, \cdots, \beta_k$, and functions $\varphi_1, \cdots, \varphi_r$. We suppose that the parameters are known and are restricted to some region B in a k-dimensional space $\{\beta_j\}$. The functions φ_r are chosen perhaps from some family of functions $\{\varphi_j\}$. The system auxiliary to system (23.1) is taken to be

$$(23.2) \qquad \frac{dx_i}{dt} = X_i(x_1, \cdots, x_n, t, \quad \beta_1^0, \cdots, \beta_k^0, \quad \varphi_1^0, \cdots, \varphi_r^0).$$

The auxiliary system is obtained from system (23.1) by taking some fixed values $\{\beta_j^0\}$ and $\{\varphi_j^0\}$. Thus we can consider the system as obtained by the variation of parameters given by

(23.3) $R_i = X_i(x, t, \beta, \varphi) - X_i(x, t, \beta^0, \varphi^0)$.

The actual form of the functions R_i depends on the type of function we admit for the functions X_i of β and φ, and also on the property of the family of functions φ and the region in which we allow the parameters β to vary.*

The general problem has now been described. We attempt to construct a Lyapunov function v that depends not only on x and t, but also on φ and β, the definition being valid for all functions φ in the given family and all values of β in the region admitted. If a function $v(x, t, \beta, \varphi)$ can be constructed so that it has the necessary properties along trajectories of system (23.1), the stability problem is solved. This method of constructing a Lyapunov function is clearly more flexible for investigating problems of stability than the method expounded in Chapter 4, where the entire parametric family (23.1) had to be subjected, so to speak, to the action of a single Lyapunov function. Unfortunately, there is no effective general rule for constructing a Lyapunov function $v(x, t, \beta, \varphi)$ in the context just discussed. Nevertheless, the method has been applied to a variety of special cases and classes of equations, many of which have particular interest in practical cases, and has been effective in obtaining stability criteria. In this way the method of constructing Lyapunov functions with parameters has made it possible to attack very broad classes of nonlinear systems.

One approach to constructing Lyapunov functions that is always useful, even when the equations of perturbed motion involve complicated functions R_i, is to bring into play the physical laws that the system under consideration is known to obey. Četaev and his co-workers have for some time been investigating this approach for mechanical systems [30], [32]–[36]. The method is as follows: We write down known integrals of the motion and construct combinations of these integrals. If some combination turns out to be a positive-definite function, then this function satisfies the conditions of Lyapunov's theorem on stability, since the derivative dv/dt of a combination of integrals of the equations of perturbed motion is identically zero.

This method has been successful for a number of stability problems involving mechanical systems. See, for example, the works of Rumyancev [166], [167], [169], who studied the stability of motion of a rigid body with a fixed point.

We remark further that the parametric method of constructing a Lyapunov function v may be especially valuable in conjunction with the generalizations of Lyapunov's second method that were given in Chapter 3. Indeed, it is seldom possible to construct a single function $v(x, t, \beta, \varphi)$ that has all the properties required by the stability or instability theorem in question for all values of β and φ that arise in a particular problem (23.1).

---

* A detailed statement of the stability problem under variation of parameters that is basically identical with the statement just given appears in the works of Kuz'min [102]. See also the work of Aminov [8].

In this chapter we shall give some examples that are illustrative of the parametric method; we cannot expect to give a complete recipe, since no such recipe exists at the present time. Furthermore, Lyapunov's method is not the only method for investigating problems of stability. The theory of differential equations includes other methods (for example, methods connected with estimating integrals and studying characteristic curves and hyperplanes). Perhaps a combination of Lyapunov's second method and these other methods will be needed to solve a particular problem. In this connection, see the works of Erugin [49]–[53] and of his pupils, Zubov [198], [199], Pliss [150]–[152], Tuzov [179], and Skačkov [172]. (There is a large literature on the subject; these references are selected only because they are connected with the methods of this book.)

24. Stability for Persistent Disturbances that are Bounded in the Mean

It is frequently useful to modify a function $v(x, t)$ that is connected with the unperturbed system by multiplying v by a suitable function $\psi(x, t)$, or, in particular, by a function $\psi(t)$. This method was invented by Četaev [36] and extended in several papers. See, for example, Razumihin [155] and Lebedev [106], where the linear system $dx/dt = x P(t)$ is considered. Here we use this method to study the stability of motion for persistent disturbances, a problem also studied in the following: Artem'ev [11], Goršin [63], [64], Dubošin [40], Erugin [54], Malkin [119], [124], Massera [130], and Četaev [36].

The problem of stability for persistent disturbances is stated as follows by Malkin [124]. We consider the two systems of equations

$$(24.1) \qquad \frac{dx_i}{dt} = X_i(x_1, \cdots, x_n, t) \qquad (i = 1, \cdots, n)$$

and

$$(24.2) \qquad \frac{dx_i}{dt} = X_i(x_1, \cdots, x_n, t) + R_i(x_1, \cdots, x_n, t) \qquad (i = 1, \cdots, n) .$$

The functions R_i are to be regarded in this section as the persistent disturbances. The precise values of these functions R_i are generally not known, and it need not be true that they reduce to zero at the point $x = 0$.

The unperturbed motion $x = 0$ is the solution of equations (24.1). This solution is said to be stable for persistent disturbances if for every positive number ε, however small, there are two positive numbers $\eta_1(\varepsilon)$ and $\eta_2(\varepsilon)$ such that whenever the initial values x_{i0} and t_0 satisfy the inequalities

$$|x_{i0}| < \eta_1(\varepsilon), \quad t_0 > 0 ,$$

the relation $|x_i| < \varepsilon$ will be valid for all $t \geq t_0$, provided that the disturbances R_i satisfy the inequalities

$$|R_i(x_1, \cdots, x_n, t)| \leq \eta_2(\varepsilon) .$$

Thus the only requirement on the functions R_i is that they are bounded in modulus according to this last relation for all t such that $t \geqq t_0$, for sufficiently small values of x, $|x_i| < \varepsilon$. Goršin [63], [64] and Malkin [125] showed that if the null solution $x = 0$ is asymptotically stable uniformly with respect to x_0 and t_0, it is also stable under persistent disturbances if the functions X_i have bounded derivatives $\partial X_i / \partial x_j$. Kurcveǐl' [97] and Massera [131] showed that if the functions $X_i(x, t)$ are periodic, the same conclusion is valid if the functions X_i are assumed to be merely continuous.

The requirement that the functions R_i be small for all values of the time is not a happy one. In some cases, it is more natural to ask only that the intervals of time $t > t_0$ during which R_i has a very large value should be extremely short. A function R_i with this property is called *bounded in the mean*. Heuristically considered, this does seem to be a sufficient condition for stability.

The ideas of this section go back to Germaǐdze [59], who considered first-order stability. Extensions of Germaǐdze's ideas are contained in [60], a joint work of Germaǐdze and the author.

We now define stability for persistent disturbances that are bounded in the mean.

DEFINITION 24.1. *The null solution* $x = 0$ *of the system* (24.1) *is called stable for persistent disturbances that are bounded in the mean if for every positive* ε ($\varepsilon > 0$) *and positive* T ($T > 0$), *there are two positive numbers* δ, η ($\delta > 0$, $\eta > 0$) *such that whenever the continuous function* $\varphi(t)$ *satisfies the relations*

$$(24.3) \qquad \int_t^{t+T} \varphi(s)\, ds < \eta \,, \quad |R_i(x, t)| \leqq \varphi(t) \quad for \ \|x\|_2 < \varepsilon \,,$$

every solution $x(x_0, t_0, t)$ *of equation* (24.2) *which has initial values* $\|x_0\|_2 < \delta$ *will satisfy* $\|x(x_0, t_0, t)\| < \varepsilon$ *for all* $t \geqq t_0$.

This definition is a generalization of Malkin's definition [124, p. 129] in the sense that a solution which is stable for persistent disturbances that are bounded in the mean is obviously *a fortiori* stable for persistent disturbances that are everywhere bounded (the number η of definition 24.1 is related to the number η_2 of Malkin's definition by the relation $\eta_2 T = \eta$).

We are assuming that the functions X_i are defined and continuous for $\|x\| < H$ and satisfy a Lipschitz condition there (with respect to x). See equation (1.7), with $L = L_H$.

It is not necessary to make any assumptions whatever concerning the smoothness or continuity of the functions R_i. We need only suppose that a trajectory $x(x_0, t_0, t)$ can be defined for $t > t_0$ for every $t_0 \geqq 0$ and every point $x_0 \in H$, and that this trajectory can be extended for all $t > t_0$ for which the values of the function $x(x_0, t_0, t)$ lie in the region $\|x\| < H$. It is required that the function $x(x_0, t_0, t)$ be a solution of equation (24.2), perhaps in some extended sense, but this must be a sense in which the relations

$$(24.4) \qquad \left(\frac{dx(x_0, t_0, t)}{dt} \right)_{dt = +0} = X_i + R_i$$

hold.

THEOREM 24.1. *Suppose the null solution $x = 0$ is asymptotically stable uniformly in t_0 and x_0 in the sense of definition 5.1 (p. 26). Then stability also holds for persistent disturbances that are bounded in the mean.*

Remark. If the functions $X_i(x, t)$ are periodic in the time t, or if they have the form $X_i(x)$ (that is, are independent of the time t), then uniform stability is an obvious consequence of asymptotic stability. Theorem 24.1 therefore asserts in this case that mere asymptotic stability carries with it stability for persistent disturbances that are bounded in the mean.

PROOF. By theorem 5.1, the hypotheses of theorem 24.1 show that in some neighborhood of the point $x = 0$ there is a function $v(x, t)$ that satisfies the relations (24.5)–(24.7):

$$(24.5) \qquad w(x) \leqq v(x, t) \leqq W(x) ,$$

where $w(x) > 0$ for $|| x || \neq 0$, and $W(0) = 0$;

$$(24.6) \qquad \left(\frac{dv}{dt} \right)_{(24.1)} \leqq -w_1(x) ,$$

where $w_1(x) > 0$ for $|| x || \neq 0$; and

$$(24.7) \qquad \left| \frac{dv}{dx_i} \right| < N \qquad (i = 1, \cdots, n)$$

for

$$(24.8) \qquad || x || < H_1 \qquad (H_1 = \text{const.}) .$$

Let a positive number $\varepsilon > 0$ be given. There is no loss of generality in assuming $\varepsilon < H_1$. The left inequality of (24.5), i.e., the positive-definiteness of the function $v(x, t)$, shows that

$$(24.9) \qquad \inf [v(x, t) \quad \text{for} \quad || x || = \varepsilon] \geqq \min [w(x) \quad \text{for} \quad || x || = \varepsilon] = 2\alpha > 0 ,$$

and by (24.7) we have the inequality

$$(24.10) \qquad \sup \left[v(x, t) \quad \text{for} \quad || x || \leqq \delta = \frac{\alpha}{nNe^2} \right] \leqq \frac{\alpha}{e^2} .$$

Now let c, c_3 be defined by

$$c = \inf \left[\frac{w(x)}{|| x ||} \right], \qquad c_3 = \inf \left[\frac{w_1(x)}{|| x ||} \right] \qquad \text{for } \delta \leqq || x || \leqq \varepsilon .$$

Then the inequalities

$$(24.11) \qquad c || x || \leqq v(x, t) \leqq nN || x || , \qquad \left(\frac{dv}{dt} \right)_{(24.1)} \leqq -c_3 || x ||$$

hold in the region

$$(24.12) \qquad \delta \leqq || x || \leqq \varepsilon .$$

In the region (24.12), the function $V(x, t) = v(x, t)/c_3$ satisfies the inequalities

(24.13) $$c_2 \| x \| \leq V(x, t) \leq c_1 \| x \| ,$$

(24.14) $$\left(\frac{dV}{dt} \right)_{(24.1)} \leq -\| x \| ,$$

(24.15) $$\left| \frac{\partial V}{\partial x_i} \right| < \frac{c_1}{n} \qquad (i = 1, \cdots, n) ,$$

where c_1 and c_2 are positive constants. From the relation $R_i \leq \varphi(t)$ [see (24.3)] we conclude that the functions R_i satisfy the inequality

(24.16) $$| R_i(x, t) | \leq \frac{\varphi(t)}{\delta} \| x \| \qquad (i = 1, \cdots, n)$$

in the region (24.12). We assert that we can choose the function $\beta(t)$ in such a way that the function $f(x, t) = V \exp [\beta(t)]$ may be used to establish the stability of the null solution $x = 0$ for persistent disturbances that satisfy the estimate (24.3). The computation of the derivative of the function $f(x, t)$ along trajectories of the system (24.2) is as follows:

(24.17) $$\lim_{\Delta t \to +0} \frac{\Delta f(x(t), t)}{\Delta t}$$

$$= f(x(t), t) \left[\beta'(t) + \frac{1}{V} \left(\frac{dV}{dt} \right)_{(24.1)} + \frac{1}{V} \sum_{i=1}^{n} \frac{\partial V}{\partial x_i} R_i \right]$$

$$\leq f(x(t), t) \left[\beta'(t) - \frac{1}{c_1} + \frac{c_1 \varphi(t)}{c_2} \right].$$

Let q be a number between 0 and 1: $0 < q < 1$. We seek to determine a nonnegative function $\psi(t)$ such that the relation

(24.18) $$\int_{kT}^{(k+1)T} \psi(t) \, dt = \int_{kT}^{(k+1)T} \left[\frac{1-q}{c_1} - \frac{c_1 \varphi(t)}{c_2 \delta} \right] dt \qquad (k = 0, 1, 2, \cdots)$$

holds. If we choose the number η of relation (24.3) to have the value

(24.19) $$\eta = \frac{T(1-q)c_2 \delta}{c_1^2} ,$$

the value of the integral (24.18) is nonnegative. Thus a nonnegative $\psi(t)$ will exist [$\psi(t) \geq 0$]. We now define the function $\beta(t)$ by the relation

(24.20) $$\beta(t) = \int_0^t \left[-\psi(\xi) - \frac{c_1 \varphi(\xi)}{c_2 \delta} + \frac{1-q}{c_1} \right] d\xi .$$

Substituting (24.20) in (24.17), we obtain

(24.21) $$\left[\lim_{\Delta t \to +0} \frac{\Delta f}{\Delta t} \right]_{(24.2)} \leq f(x, t) \left[-\psi(\xi) - \frac{q}{c_1} \right] \leq - \frac{q}{c_1} f .$$

On the other hand, conditions (24.18) and (24.20) show that $\beta(kT) = 0$ for $k = 1, 2, \cdots$, and consequently $| \beta(t) | \leq 3(1-q)T/c_1$.

If we now choose q so that $(1 - q)T/c_1 < 1/3$, it follows that the inequalities

(24.22) $$Ve^{-1} \leq f \leq Ve$$

hold in the region (24.12).

Thus the function f that we have constructed satisfies condition (24.21), and in view of inequalities (24.9), (24.10), and (24.22), it also satisfies the inequality

(24.23) $$\sup [f(x, t) \quad \text{for} \ || x || \leq \delta] < \inf [f(x, t) \quad \text{for} \ || x || = \varepsilon] .$$

Now let us consider a trajectory $x(x_0, t_0, t)$ of the system (24.1) when x_0 is restricted to be less than δ: $|| x_0 || < \delta$. By the inequalities (24.21) and (24.23), the trajectory $x(x_0, t_0, t)$ remains in the region $|| x || < \varepsilon$, which demonstrates the theorem.

Remark. By introducing a slight complication in the above method of constructing the function $f(x, t)$ it is possible to demonstrate a stronger assertion. To state this assertion, suppose γ is a positive number less than ε $(0 < \gamma < \varepsilon)$, where ε is the number mentioned above. The numbers c and c_3 are defined by the relations

$$c = \inf \left[\frac{w(x)}{|| x ||} \right], \qquad c_3 = \inf \left[\frac{w_1(x)}{|| x ||} \right] \qquad \text{for} \ || x || > \gamma_1 ,$$

where γ_1 is chosen so that $W(\gamma_1)$ is less than $w(\gamma)$. (Note that the range over which the infimum was taken above was $|| x || \geq \delta$.) By methods of construction and calculation exactly analogous to those above, we can find a function f that satisfies inequality (24.23) and also inequality (24.21), the latter (naturally for appropriate constants) everywhere in the region $\gamma_1 \leq || x || \leq \varepsilon$. Thus we can show with the help of this function $f(x, t)$ that the trajectories $x(x_0, t_0, t)$ of the system (24.1) not only remain in the region $|| x || < \varepsilon$ for all $t \geq t_0$, but also satisfy the relation

$$\lim_{t \to \infty} \sup || x(x_0, t_0, t) || \leq \gamma$$

whenever the mean value of the persistent disturbances is bounded by a sufficiently small constant.

We emphasize also that the way we have constructed f for a particular system (24.1) and particular functions $V(x, t)$ and $\varphi(t)$ allows us not only to establish stability for persistent disturbances but also to obtain useful concrete estimates that are exceedingly good if the function $v(x, t)$ is properly chosen. (The better the choice of the function v, the better the estimates.)

We next show how to construct a function $f(x, t)$ to solve a special problem. It will be recalled that terms containing harmonics of high order are frequently neglected in a general analysis of the stability of solutions of periodic systems. The following chain of theorems is a rigorous foundation for such a procedure.

We return to system (24.1) and assume that the functions X_i satisfy the conditions previously imposed. We also assume that the functions X_i are uniformly continuous in t, and that the functions $R_i(x, t)$ are bounded and satisfy a Lipschitz condition with respect to x_j; that is,

$$(24.24) \qquad\qquad |R_i(x, t)| \leq K,$$
$$(24.25) \qquad\qquad |R_i(x'', t) - R_i(x' t)| \leq L \| x'' - x' \|.$$

Moreover, we shall suppose that the functions $R_i(x, t)$ are periodic in the time t with period $\tau > 0$, and that their Fourier series contain no constant term (any possible constant term may be thought of as assimilated into the function X_i). In other words, we suppose that the functions R_i satisfy the inequality

$$(24.26) \qquad\qquad \int_t^{t+\tau} R_i(\tilde{x}, \xi)\, d\xi = 0 \qquad (i = 1, \cdots, n)$$

for every fixed* value $x = \tilde{x}$.

THEOREM 24.2. *If the null solution $x = 0$ of the system of equations (24.1) is asymptotically stable uniformly in x_0 and t_0, then for arbitrary positive ε ($\varepsilon > 0$), there exist two positive numbers $\delta > 0$, $\tau_0 > 0$ with the following properties. Whenever the initial value $\| x_0 \|$ is less than δ, and whenever relation (24.26) is satisfied for a single value of τ, $\tau \leq \tau_0$, the solution $x(x_0, t_0, t)$ of equation (24.2) will satisfy the inequality $\| x(x_0, t_0, t) \| < \varepsilon$.*

Remark. Theorem 24.2 says, in effect, that if the null solution of a system is uniformly asymptotically stable, sufficiently fast vibrations of even large amplitude cannot greatly change this stability. It should be noted further that the two assertions in the remark following theorem 24.1 are valid (with necessary changes) for theorem 24.2. It is especially interesting to note that if the functions $X_i(x, t)$ in theorems 24.1 and 24.2 are periodic in the time t, the theorems are valid if we suppose only that the functions X_i are continuous. Indeed, according to the results obtained in the works of Kurcveĭl' [97] and Massera [131], the functions $v(x, t)$ used in the proof of theorems 24.1 and 24.2 to construct the function $f(x, t)$ do exist under the hypothesis that the functions X_i are only continuous.

PROOF. Following the proof of theorem 24.1, we consider the Lyapunov function $f(x, t) = V(x, t) \exp \beta(t)$, where the function $V(x, t)$ satisfies the estimates (24.13)–(24.15). The proof will be made by contradiction. Thus we assume that no matter how small a number γ_0 is chosen, we can find a system (24.2) which does satisfy the conditions of the theorem, and for which a trajectory $x(x_0, t_0, t)$ that lies initially ($t_0 \geq 0$) on the surface $\| x_0 \| = \delta$ lies for $t_0 \leq t < t_1$ entirely in the region (24.12), and is outside the region $\| x \| = \varepsilon$ for $t = t_1 > t_0$. The first step is to compute the right derivative df/dt along a trajectory of the system (24.2):

---

* The only way in which the periodicity of the functions R_i is used is the assumption (24.26). Thus when we refer to functions R_i in subsequent pages, we have in mind arbitrary functions that satisfy (24.24), (24.25), and (24.26).

$$\left(\lim_{\Delta t \to +0} \frac{\Delta f}{\Delta t}\right)_{(24.2)} = e^{\beta(t)}\left[V\beta'(t) + \left(\frac{dV}{dt}\right)_{(24.1)} + \sum_{i=1}^{n}\frac{\partial V}{\partial x_i}R_i\right].$$

From estimates (24.13)–(24.15) we obtain the inequality

$$\left(\frac{df}{dt}\right)_{(24.2),\,dt=+0} \leq f(x,t)\left[\beta'(t) - \frac{1}{c_1} + \frac{1}{V}\sum_{i=1}^{n}\frac{\partial V}{\partial x_i}R_i\right].$$

We must now select a function f so that on the trajectory $x(x_0, t_0, t)$ that we are considering we can apply theorem 17.1. Thus our problem is to define the function $\beta(t)$ so that the function $f(x,t)$ is positive-definite and admits an infinitely small upper bound, and so that its derivative $(df/dt)_{dt=+0}$ is negative-definite in the region (24.12) along the trajectory $x(x_0, t_0, t)$ that we are considering. We choose the functions $\omega(t)$ and $\beta(t)$ to satisfy the following relations:

$$(24.27) \qquad \int_{\theta_k}^{\theta_{k+1}} \omega(\xi)\, d\xi = \frac{1-q}{c_1}\tau$$

$$- \int_{\theta_k}^{\theta_{k+1}} \frac{1}{V}\left[\sum_{i=1}^{n}\frac{\partial V(x(x_0,t_0,t),t)}{\partial x_i}R_i(x(x_0,t_0,t),t)\right]dt\,,$$

$$(24.28) \quad \beta(t) = \frac{1-q}{c_1}(t - t_0)$$

$$- \int_{t_0}^{t}\left[\sum_{i=1}^{n}\frac{\partial V(x(x_0,t_0,\xi),\xi)}{\partial x_i}\frac{1}{V}R_i(x(x_0,t_0,\xi),\xi)\right]d\xi - \int_{t_0}^{t}\omega(\xi)\,d\xi\,,$$

where $\theta_k = t_0 + k\tau$, $k = 0, 1, \cdots$, $t \geq t_0$, and where $0 < q < 1$. We must estimate the value of the integral in the right member of equation (24.27). By condition (24.26), we have

$$(24.29) \qquad \int_{\theta_k}^{\theta_{k+1}} \frac{1}{V}\left[\sum_{i=1}^{n}\frac{\partial V(x(x_0,t_0,t),t)}{\partial x_i}R_i(x(x_0,t_0,t),t)\right]dt$$

$$= \int_{\theta_k}^{\theta_{k+1}}\left(\sum_{i=1}^{n}\left[\frac{\partial V(x(x_0,t_0,t),t)}{\partial x_i}\frac{R_i(x(x_0,t_0,t),t)}{V(x(x_0,t_0,t),t)}\right.\right.$$
$$\left.\left. - \frac{\partial V(x_k,\theta_k)}{\partial x_i}\frac{R_i(x_k,t)}{V(x_k,\theta_k)}\right]\right)dt\,,$$

where x_k is $x(x_0, t_0, \theta_k)$.

We now apply the theorem of the mean to the right member of equation (24.29), and thus estimate for the left member of this equation the value $\tau N n$, where

$$(24.30) \qquad N = \sup\left|\frac{\partial V(x(t),t)}{\partial x_i}\frac{R_i(x(t),t)}{V(x(t),t)} - \frac{\partial V(x_k,\theta_k)}{\partial x_i}\frac{R_i(x_k,t)}{V(x_k,\theta_k)}\right|,$$

$$\theta_k \leq t \leq \theta_{k+1}\,.$$

Since the function V is continuous with bounded derivatives of all orders with respect to all its arguments (theorem 5.1), we can conclude from estimates (24.24) and (24.25) that the relation

$$(24.31) \qquad\qquad\qquad N \leq Q\tau$$

holds ($Q = $ const.). [We omit the details of the derivation of (24.31), which is straightforward.]

The estimate

(24.32) $$\int_{\theta_k}^{\theta_{k+1}} \frac{1}{V}\left[\sum_{i=1}^{n} \frac{\partial V(x(t), t)}{\partial x_i} R_i(x(t), t)\right] dt \leq P\tau^2 \qquad (P = \text{const.})$$

is an immediate consequence of (24.31). We now subject the number τ ($\tau > 0$) to the requirement

(24.33) $$\tau < \frac{1 - q}{c_1 P} \;.$$

If τ satisfies (24.33), the right member of equation (24.27) is positive. Thus the function $\omega(t)$ can be taken to be nonnegative if (24.33) holds. We can further assume that the function $\omega(t)$ is nonnegative, $\omega(t) \geq 0$, for $t \geq t_0$. We note from (24.28) that the function $\beta(t)$ satisfies the equalities $\beta(\theta_k) = 0$ ($k = 1, 2, \cdots$); thus we can assert that $|\beta(t)| < P_1\tau$, $P_1 = $ const. If the number $\tau > 0$ is also bounded by the relation $\tau < 1/P_1$, then we see that $|\beta(t)|$ is less than 1: $|\beta(t)| < 1$. Thus the function $f(x, t)$ satisfies estimates (24.21), (24.23) when $\gamma > 0$ is bounded in this manner. This conclusion contradicts the supposition that the trajectory $x(x_0, t_0, t)$ leaves the region $\|x\| \leq \varepsilon$ for $t > t_0$. Finally, since the numbers P, q, which are used to obtain the bounds for γ, depend only on the properties of the functions X_i and on the constants K, L of relations (24.24), (24.25), and are independent of the functions $R_i(x, t)$, the proof is complete.

Now we describe methods for constructing a Lyapunov function for the linear system of equations

(24.34) $$\frac{dx_i}{dt} = \sum_{j=1}^{n} p_{ij}(t)x_j \qquad (i = 1, \cdots, n) \;.$$

These methods are detailed by Razumihin [155] and Lebedev [106]. We assume that the roots of the characteristic equation

(24.35) $$\det [p_{ij}(t) - \lambda\delta_{ij}]_1^n = 0$$

satisfy the relation

(24.36) $$\text{Re } \lambda_i < -\gamma \qquad (i = 1, \cdots, n) \;,$$

where $\gamma > 0$ is a positive constant, for each fixed value of t. Lyapunov's theorem then assures us that there is a positive-definite function

(24.37) $$v(x, t) = \sum_{i,j=1}^{n} a_{ij}(t)x_i x_j$$

which satisfies the relation

(24.38) $$\sum_{i=1}^{n} \frac{\partial v}{\partial x_i}\left[\sum_{j=1}^{n} p_{ij}(t)x_j\right] = -\sum_{i=1}^{n} x_i^2 \;.$$

We assume that the coefficients $p_{ij}(t)$ are bounded differentiable functions of the time t. The coefficients $a_{ij}(t)$ must then also be bounded differentiable functions of t. Following our general plan, we take a Lyapunov function $f(x, t)$ in the form $f(x, t) = v(x, t) \exp \beta(t)$. The value of the derivative df/dt of this function along a trajectory of the system (24.34) is, according to (24.38),

$$\frac{df}{dt} = -\sum_{i=1}^{n} x_i^2 + \sum_{i,j=1}^{n} \left[\frac{da_{ij}(t)}{dt} + \frac{d\beta(t)}{dt} a_{ij}(t) \right] x_i x_j .$$

We consider the quadratic form

$$w(x, t) = -\sum_{i=1}^{n} x_i^2 + \sum_{i,j=1}^{n} \left[\frac{da_{ij}(t)}{dt} + \frac{d\beta(t)}{dt} a_{ij}(t) \right] x_i x_j .$$

Since $w(x, t)$ is a quadratic form, a general theorem assures us that this form satisfies the inequality

$$w(x, t) \leqq \rho_{\max} \sum_{i=1}^{n} x_i^2 ,$$

where $\rho_{\max}$ is the greatest root of the determinantal equation

(24.39) $\det [c_{ij}(t) - \rho \delta_{ij}]_1^n = 0$ $[c_{ij}(t) = -\delta_{ij} + a'_{ij}(t) + \beta'(t) a_{ij}(t)]$.

Thus, to assure that the function $w(x, t)$ will be negative-definite, it is sufficient to choose $\beta(t)$ in such a way that the roots of equation (24.39) satisfy the relation

(24.40) $$\rho_{\max} < -\alpha ,$$

where $\alpha > 0$ is a positive constant. Condition (24.40) gives an upper bound for the function $\beta'(t)$, since the form $v(x, t)$ is positive-definite because of the way the function v (24.37) was chosen. Let us call this bound $\bar{\beta}(t)$. On the other hand, if the function f is to be positive-definite and the function df/dt negative-definite, we must have

(24.41) $$\beta(t) > N > -\infty \qquad (N = \text{const.}) .$$

Relation (24.41) requires only that we take

(24.42) $$\int_0^t \bar{\beta}(t) dt = N_1 \geqq -\infty .$$

Thus a criterion sufficient for asymptotic stability of the solutions of the system (24.34) is that relation (24.42) be satisfied by $\bar{\beta}(t)$ [defined as the greatest number for which the roots of equation (24.39) satisfy inequality (24.40)]. An effective estimate of $\bar{\beta}(t)$ can be obtained from the following considerations.

Let $b_{ij}(t) = c_{ij}(t) - \beta'(t) a_{ij}(t)$. If $\lambda_{\max}$ is the greatest value of λ that satisfies the equation.

(24.43) $$\det [c_{ij}(t) - \lambda a_{ij}(t)] = 0 ,$$

the form $\sum_{i,j=1}^{n} c_{ij}(t) x_i x_j$ will be negative-definite if $\lambda_{\max} \leqq -\alpha$. Thus, if $\rho_{\max}$ is the greatest value of ρ that satisfies the equation

(24.44) $\det [b_{ij}(t) - \rho a_{ij}(t)] = 0$,

a sufficient condition for asymptotic stability of the solutions of the system (24.34) is that for some constant M independent of t, the relation

(24.45) $$\int_0^\infty (\rho_{\max} + \alpha) \, dt < M < \infty$$

hold. Indeed, if this relation is satisfied, we can take $\bar{\beta}(t) = -\rho_{\max}(t) - \alpha$. Thus we have shown that we can find an effective bound $\bar{\beta}(t)$ by taking the greatest root ρ of the equation

$$\det \left[\frac{da_{ij}}{dt} - (1 - \alpha)\rho a_{ij} - \delta_{ij} \right]_1^n = 0 ,$$

with reversed sign.

25. Criteria for Asymptotic Stability in the Large for Some Nonlinear Systems

In this section and the following ones we consider a method for constructing a Lyapunov function $v(x, t, \beta, \varphi)$ that was originated in the work of Lur'e and Postnikov [112] and afterward used widely in many places. (See, for example, [15], [27], [41], [47], [51], [82], [83], [85], [86], [108]–[110], [113], [120], [122].)

In considering a servomechanism it is necessary to study the behavior of the solutions of a system of differential equations of the following form [5], [113]:

(25.1) $$\frac{dx_i}{dt} = d_{i1}x_1 + \cdots + d_{in}x_n + \beta_{i1}\varphi_{i1}(x_1) + \cdots + \beta_{in}\varphi_{in}(x_n)$$

$$(i = 1, \cdots, n) ,$$

where d_{ij}, β_{ij} are constants, and $\varphi_{ij}(x_j)$ are nonlinear functions.[*]

In the stability problem for system (25.1), which describes servomechanisms, it is desirable to treat the possibility of a rather large initial perturbation $(x_{10}, x_{20}, \cdots, x_{n0})$, i.e., to consider equation (25.1) in a sufficiently large region of phase space. Thus a satisfactory solution to the problem cannot be obtained by investigating the linearized version of (25.1), since there is in fact no linear system whose right member would differ from the right member of system (25.1) by a sufficiently small value in all points of the region in which the initial perturbation $(x_{10}, x_{20}, \cdots, x_{n0})$ is being considered.

An effective solution of the necessary stability problems is possible only if we construct the Lyapunov function v in such a way that the nonlinearity of the functions φ_{ij} is properly mirrored. The idea of Lur'e and Postnikov for constructing a Lyapunov function $v(x, t, \beta, \varphi)$ for system (25.1) is to allow the function to depend both on integrals of the functions φ and on a quad-

---

[*] Usually the functions φ_i depend on linear combinations $\sigma_i = \xi_{i1}x_1 + \cdots + \xi_{in}x_n$, where $\xi_{ij} = \text{const.}$, and a linear change of variables will bring this to the form (25.1). We confine our attention to this form.

ratic form in the arguments $x_1, \cdots, x_n$. The precise form needed is the linear combination

$$(25.2) \qquad v(x, t, \beta, \varphi) = \sum_{i,j=1}^{n} a_{ij}x_ix_j + \sum_{i,j=1}^{n} \alpha_{ij}\int_0^{x_j}\varphi_{ij}(\xi)\,d\xi \ .$$

We shall see that it is possible to give a formula for the derivative dv/dt along a trajectory of (25.1) and to give conditions that this derivative be positive-definite or semidefinite. It is also possible to give corresponding conditions concerning the definiteness or semidefiniteness of the function $v(x, t, \beta, \varphi)$. We shall then have criteria for the stability or instability of system (25.1). These criteria depend essentially on the rule by which the form $\sum a_{ij}x_ix_j$ and the coefficients $c_{ij}(\beta)$ are chosen in definition (25.2). Various rules and corresponding conditions of stability are contained in the works cited in the first paragraph of this section. To classify and survey the various stability criteria, it is desirable to classify the conditions with respect to a certain reasonably small number of invariants. A description of the construction of the various functions $v(x, t, \beta, \varphi)$ is carefully set out by Aizerman. In his study of servomechanisms, he gave a very natural and interesting statement of the stability problem for systems of the type (25.1). We formulate this problem below.*

We now consider system (25.1) and the linear system

$$(25.3) \qquad \frac{dx_i}{dt} = d_{i1}x_1 + \cdots + d_{in}x_n + \beta_{i1}h_{i1}x_1 + \cdots + \beta_{in}h_{in}x_n$$

$$(i = 1, \cdots, n) \ .$$

We assume that solutions of the linear system (25.3) are asymptotically stable when the parameters h_{ij} lie in a region

$$(25.4) \qquad\qquad\qquad \underline{h}_{ij} < h_{ij} < \bar{h}_{ij} \qquad\qquad\qquad \text{(a)}$$

or in a region

$$(25.5) \qquad\qquad\qquad \underline{h}_{ij} \leqq h_{ij} \leqq \bar{h}_{ij} \ , \qquad\qquad\qquad \text{(b)}$$

where $\underline{h}_{ij}$ and $\bar{h}_{ij}$ are constants. We shall study the asymptotic stability of the null solution $x = 0$ of the nonlinear system (25.1) for arbitrary initial perturbations, when the nonlinear function $\varphi_{ij}(x_j)$ satisfy condition

$$(25.6) \qquad\qquad \underline{h}_{ij} < \frac{\varphi_{ij}(x_j)}{x_j} < \bar{h}_{ij} \qquad (x_j \neq 0) \ , \qquad\qquad \text{(a)}$$

or condition

$$(25.7) \qquad\qquad \underline{h}_{ij} \leqq \frac{\varphi_{ij}(x_j)}{x_j} \leqq \bar{h}_{ij} \qquad (x_j \neq 0) \ . \qquad\qquad \text{(b)}$$

---

* Aizerman [3] formulated the problem for the case in which system (25.1) contains a single nonlinear function $\varphi_{ij}(x_j)$ and in which, moreover, there is only a single line, that is, only a single i. Our present formulation is an obvious generalization of the problem to a system (25.1) containing several nonlinear functions $\varphi_{ij}(x_j)$.

Nonlinear continuous functions φ_{ij} that satisfy (25.6) are called functions of class (a), and functions φ_{ij} that satisfy (25.7) are called functions of class (b). The stability problem that corresponds to the one formulated at the beginning of this chapter may be stated as follows:

Find conditions for the stability of system (25.1) for parametric perturbations defined by functions φ_{ij} of class (a) [or class (b)]. Aĭzerman's studies of this problem appear in a large number of articles (see references at the beginning of this section). A complete solution of the problem is available only in the case $n = 2$. For $n \geq 3$, we know only that if the coefficients d_{ij} are connected by certain relations, the problem has a positive solution. (See, for example, [15], [27], [152], [170], [179].)* Here we restrict ourselves to consideration of the case $n = 2$.

Aĭzerman's problem for a system of two equations with a single nonlinear function, that is, a system of the form

$$(25.8) \qquad \frac{dx}{dt} = ax + by + f(x) , \qquad \frac{dy}{dt} = cx + dy ,$$

or

$$(25.9) \qquad \frac{dx}{dt} = ax + by , \qquad \frac{dy}{dt} = cx + dy + f(x) ,$$

was studied in detail by Erugin [49]-[51], who gave a number of cases in which system (25.8) has a positive solution for problems (a) and (b). For system (25.9) it has been shown [50] that Aĭzerman's problem (a) [and therefore *a fortiori* problem (b)] has a positive solution. By constructing an explicit Lyapunov function v for systems (25.8) and (25.9), Malkin [122] showed that both systems have a positive solution for problem (b). Later, it was shown that problem (a) for system (25.8) has a negative solution [81], so that the fundamental necessity of distinguishing between problems (a) and (b) is clear.

Here we consider the possible types of systems of two equations with two nonlinear functions. For such systems (treated in [81], [83]), we consider only problem (b), since, as we have just noted, problem (a) has a negative solution for system (25.8) even if there is a single nonlinear function.

We consider the system of equations

$$(25.10) \qquad \frac{dx}{dt} = f_1(x) + ay , \qquad \frac{dy}{dt} = f_2(x) + by ,$$

where $f_1(x)$ and $f_2(x)$ are continuous functions such that $f_1(0) = f_2(0) = 0$, and where a and b are constants.

The system (25.3) of linear equations that corresponds to the nonlinear system (25.10) has the form

$$(25.11) \qquad \frac{dx}{dt} = h_1 x + ay , \qquad \frac{dy}{dt} = h_2 x + by .$$

---

* Aĭzerman's problem was studied in detail for $n = 3$ by Pliss [152]. See also [27a], [135a], [141c].

A necessary and sufficient condition for asymptotic stability of system (25.11) is the familiar requirement that the roots λ_1 and λ_2 of the characteristic equation

$$(25.12) \qquad \det \begin{bmatrix} h_1 - \lambda & a \\ h_2 & b - \lambda \end{bmatrix} = 0$$

satisfy the inequality

$$(25.13) \qquad \operatorname{Re} \lambda_i < -\delta \qquad (i = 1, 2).$$

The Routh-Hurwitz condition [60a, p. 231] that guarantees (25.13) is the set of inequalities

$$(25.14) \qquad h_1 + b < -\gamma, \qquad h_1 b - h_2 a > \gamma,$$

where $\gamma > 0$ is a suitable positive constant. Thus a solution of problem (b) for system (25.10) will be obtained if we find sufficient conditions for the asymptotic stability of the null solution $x = y = 0$ of this system under the conditions

$$(25.15) \qquad \frac{f_1(x)}{x} + b < -\gamma, \qquad \frac{f_1(x)}{x} b - \frac{f_2(x)}{x} a > \gamma \qquad \text{for } x \neq 0,$$

where $\gamma > 0$ is a certain positive constant.

THEOREM 25.1. *If conditions* (25.15) *are satisfied, the null solution* $x = y = 0$ *of system* (25.10) *is asymptotically stable for arbitrary initial perturbations.*

PROOF. We consider only the case $a \neq 0$, since in the case $a = 0$ the variables of system (25.10) are separated and the system can be integrated immediately. [If $a = 0$, condition (25.15) is not enough. We must require also that $f_2(x)$ lie between two linear functions, i.e., that $f_2(x)$ not grow too rapidly.]

We consider the function

$$(25.16) \qquad v(x, y) = (bx - ay)^2 + 2\int_0^x (f_1(\xi)b - f_2(\xi)a)\, d\xi.$$

The function $v(x, y)$ is positive-definite and admits an infinitely small upper bound, since from (25.15) we obtain the relation

$$(25.17) \qquad v(x, y) \geq (bx - ay)^2 + \gamma x^2.$$

If $x^2 + y^2 \to \infty$, inequality (25.17) shows that $v(x, y) \to \infty$ if our assumption $a \neq 0$ is satisfied. In other words, the function $v(x, y)$ admits an infinitely great lower bound for $x^2 + y^2 \to \infty$ (in the sense of definition 5.2, p. 29).

The value of the derivative dv/dt along a trajectory of system (25.10) is as follows:

$$(25.18) \qquad \frac{1}{2}\frac{dv}{dt} = [f_1(x) + b][f_1(x)b - f_2(x)a] \leq -\gamma^2 x^2.$$

Thus the derivative dv/dt is negative-semidefinite and has the value 0 only along the line $x = 0$. Moreover, since $dx/dt \neq 0$ for $y \neq 0$, condition (14.12) is satisfied. Thus the function $v(x, y)$ satisfies the conditions of theorem 14.1, and theorem 25.1 is proved.

Theorem 14.1, which is used in the proof of theorem 25.1, was proved only under the condition that the functions X_i (in this case f_1 and f_2) satisfy a Lipschitz condition. Thus theorem 25.1 is established only in this case. It can be shown, however, that theorem 14.1 remains valid in the more general case in which the functions X_i are continuous, even if the system of equations (1.3) does not have a unique solution. Under these weaker assumptions, the proof of theorem 14.1 becomes more complicated (such a proof appears in the works of Barbašin [15] and Šimanov [170], where special applications of the function v, theorem 14.1, are given). Moreover, if the functions $f_1(x)$ and $f_2(x)$ are merely piecewise-continuous but satisfy conditions (25.15), it is possible to establish the asymptotic stability of the solution $x = y = 0$ for arbitrary initial perturbations x_0 and y_0 of system (25.10). For this, the method and result of section 17, p. 79, are sufficient.

The remaining ways in which two nonlinear functions can appear in a system of two equations are the following:

$$(25.19) \qquad \frac{dx}{dt} = f_1(x) + ay , \qquad \frac{dy}{dt} = bx + f_2(y) ;$$

$$(25.20) \qquad \frac{dx}{dt} = f_1(x) + f_2(y) , \qquad \frac{dy}{dt} = ax + by ;$$

$$(25.21) \qquad \frac{dx}{dt} = ax + f_1(y) , \qquad \frac{dy}{dt} = f_2(x) + by .$$

Methods similar to those already expounded can be used [81], [83] to derive the following results, which we cite without proof.

If $f_1(x)$ and $f_2(x)$ are nonlinear and satisfy the conditions

$$(25.22) \qquad \frac{f_1(x)}{x} + \frac{f_2(y)}{y} < 0 , \qquad \frac{f_1(x)}{x} \frac{f_2(y)}{y} - ab > 0 ,$$

the solution $x = y = 0$ of system (25.19) is asymptotically stable for arbitrary initial perturbations. Conditions (25.22) state that the Routh-Hurwitz conditions are formally satisfied [problem (a) has a positive solution]. The situation with system (25.21) is the same under the conditions

$$(25.23) \qquad a + b < 0 , \qquad ab - \frac{f_1(x)}{x} \frac{f_2(y)}{y} > 0 \qquad \text{for } x, y \neq 0 .$$

For system (25.20) the situation is different. Indeed, the conditions

$$(25.24) \qquad \frac{f_1(x)}{x} + b < -\gamma , \qquad \frac{f_1(x)}{x} b - \frac{f_2(y)}{y} a > \gamma ,$$

which correspond to the Routh-Hurwitz conditions for linear systems, are not sufficient conditions for asymptotic stability. This was established by an example in [83]. Thus problem (b) has a negative solution. A possible additional hypothesis, which together with (25.24) guarantees the global asymptotic stability of the solution $x = y = 0$ of (25.20), is the following:

(25.25)   The function $f_1(x) + bx$ is a monotonically decreasing function of x, or in particular $f_1'(x) + b < 0$.

In this case, the global asymptotic stability of system (25.20) can be established by use of the Lyapunov function

$$v(x, y) = \frac{1}{2}(ax + by)^2 - a\int_0^y \left[f_1\left(-\frac{b}{a}\xi\right) - f_2(\xi) \right] d\xi ,$$

but we do not give a detailed proof of this assertion.

We now consider the nonlinear system

(25.26) $\dfrac{dx}{dt} = X(x, y) ,$ $\dfrac{dy}{dt} = Y(x, y) .$

This system is of general form. We assume that the functions X, Y are continuous and that the partial derivatives $\partial X/\partial x$, $\partial Y/\partial y$ exist and are continuous for all x and y. We also assume that the functions X, Y are differentiable for $x = y = 0$. We suppose further, as usual, that the relation $X(0, 0) = Y(0, 0) = 0$ is satisfied.

THEOREM 25.2.   *The solution $x = y = 0$ of the system of equations (25.26) is asymptotically stable for arbitrary initial perturbations if the conditions*

(25.27) $l(x^2 + y^2) \leq X^2(x, y) + Y^2(x, y)$ *for all x, y ,*

(25.28) $\dfrac{\partial X}{\partial x} + \dfrac{\partial Y}{\partial y} < -\beta$ *for all x, y*

are satisfied where l, β are positive constants.

PROOF.   First, we shall show that the condition

(25.29) $a_{11}a_{22} - a_{12}a_{21} > 0$

is satisfied, where

$$a_{11} = \left(\frac{\partial X}{\partial x}\right)_0 , \quad a_{22} = \left(\frac{\partial Y}{\partial y}\right)_0 , \quad a_{12} = \left(\frac{\partial X}{\partial y}\right)_0 , \quad a_{21} = \left(\frac{\partial Y}{\partial x}\right)_0 .$$

denote the partial derivatives evaluated at the point $x = y = 0$.

Indeed, since the functions X, Y are differentiable at the point $x = y = 0$, we can write

(25.30)
$$X(x, y) = a_{11}x + a_{12}y + R_1(x, y) ,$$
$$Y(x, y) = a_{21}x + a_{22}y + R_2(x, y) ,$$

where the order of the functions R_1, R_2 exceeds 1 at the point $x = y = 0$. Suppose *per contra* that

(25.31) $a_{11}a_{22} - a_{12}a_{21} = 0 .$

Then we can write

(25.32) $X^2 + Y^2 = (cx + dy)^2 + Q(x, y) ,$

where c, d are constants, since by (25.31) the discriminant of the quadratic form $(a_{11}x + a_{12}y)^2 + (a_{21}x + a_{22}y)^2$ is zero.

The order of the function $Q(x, y)$ in equation (25.32) exceeds 1 at the point $x = y = 0$. Therefore, inequality (25.27) cannot be satisfied if we move from the point $x = y = 0$ along the line $cx + dy = 0$. This contradiction establishes inequality (25.29).

Thus the linearized approximation

$$(25.33) \qquad \frac{dx}{dt} = a_{11}x + a_{12}y , \qquad \frac{dy}{dt} = a_{21}x + a_{22}y$$

to the system (25.26) at the point $x = y = 0$ satisfies the Routh-Hurwitz conditions

$$a_{11} + a_{22} < -\gamma , \qquad a_{11}a_{22} - a_{12}a_{21} > \gamma ;$$

that is, the solution $x = y = 0$ of system (25.33) is asymptotically stable. The fundamental theorem of Lyapunov [114, p. 127] now shows that the null solution $x = y = 0$ of system (25.26) is asymptotically stable for sufficiently small perturbations x_0, y_0 of the initial conditions. It must be emphasized that this conclusion is reached without any estimate of the magnitude of the initial perturbation for which stability holds.

Now let us establish the fact that the region of attraction of the initial point around $x = y = 0$ includes the entire xy-plane. We assume the opposite, and plan to derive a contradiction.

By a theorem of Erugin [49], the boundary of the region of attraction around the point $x = y = 0$ for asymptotic stability consists of complete trajectories of system (25.26). The part of this boundary contained in a circle of sufficiently large radius is a compact set, since each of the trajectories that compose the boundary is a closed set and so is their union. Thus, according to our assumption, we can mark a point x_0, y_0 lying on the boundary of the region of attraction and nearest to the origin.

We consider the trajectory $x(x_0, y_0, t)$, $y(x_0, y_0, t)$. By inequality (25.27), the point (x_0, y_0) is not a singular point of the system (25.26). In fact, by (25.27), system (25.26) has no singular points except for the origin $x = y = 0$. Therefore, x_0, y_0 is the point on the trajectory $x(x_0, y_0, t)$, $y(x_0, y_0, t)$ nearest to the origin, since the direction of the trajectory, the vector

$$(X(x_0, y_0),\ Y(x_0, y_0)) ,$$

is perpendicular to the radius vector to the point x_0, y_0.

Let us show that the trajectory $x(x_0, y_0, t)$, $y(x_0, y_0, t)$ cannot remain in a bounded portion of the xy-plane for all time $t \geq 0$. By the ω-limit set of a trajectory, we mean the set of limit points of all possible subsequences of the form

$$x(x_0, y_0, t_k),\ y(x_0, y_0, t_k),\quad t_k \to \infty .$$

It is clear that a boundary half-trajectory $x(x_0, y_0, t)$, $y(x_0, y_0, t)$, where $0 \leq t < \infty$, must have a bounded nonempty ω-limit set. The book [39, p.

390] offers a proof that the ω-limit set of a given trajectory $x(t), y(t)$ consists of entire trajectories; more generally, it is shown in the same work (p. 391) that the set of trajectories consisting of the bounded ω-limit sets in the plane must be either singular points of the system or isolated trajectories. System (25.26) has no singular point except for the point $x = y = 0$. Neither can the system have an isolated trajectory. Indeed, suppose Γ is such an isolated trajectory, and D the region bounded by the contour Γ. Then Green's Formula gives the following result:

$$\oint_\Gamma (Xdy - Ydx) = \iint_{(D)} \left(\frac{\partial X}{\partial x} + \frac{\partial Y}{\partial y} \right) dxdy = 0 .$$

But this contradicts hypothesis (25.28). Hence a boundary trajectory cannot be bounded for all time $t \geq 0$.

Now let us suppose that the trajectory $x(x_0, y_0, t)$, $y(x_0, y_0, t)$ has points lying at an arbitrary distance from the origin. Let R be an arbitrary positive number.

By assumption, we can find a point $x(x_0, y_0, t_1)$, $y(x_0, y_0, t_1)$ lying outside the circle

(25.34) $x^2 + y^2 \leq R.$

The solutions of (25.26) depend continuously on the initial values. Hence there is a value ξ, $0 < \xi < 1$, such that if $x_1 = \xi x_0$, $y_1 = \xi y_0$, the trajectory $x(x_1, y_1, t)$, $y(x_1, y_1, t)$ also contains a point $x(x_1, y_1, t_1)$, $y(x_1, y_1, t_1)$ lying outside the circle (25.34). Now $x(x_1, y_1, t) \to 0$, $y(x_1, y_1, t) \to 0$ for $t \to \infty$, since the point (x_1, y_1) lies closer to the origin than the point (x_0, y_0). Thus the point (x_1, y_1) lies in the region of attraction of the point $x = y = 0$.

We now consider the trajectories

(25.34a) $x^*(x(x_1, y_1, t_1), y(x_1, y_1, t_1), t)$, $y^*(x(x_1, t_1, t), y(x_1, y_1, t_1), t)$

of the auxiliary system

(25.35) $\dfrac{dx^*}{dt} = -Y(x^*, y^*) ,$ $\dfrac{dy^*}{dt} = X(x^*, y^*) .$

It is clear that the trajectories of system (25.35) are orthogonal to the trajectories of system (25.26). Let us consider the half-trajectory

$$x^*(x(x_1, y_1, t_1), y(x_1, y_1, t_1), t), \ y^*(x(x_1, y_1, t_1), y(x_1, y_1, t_1), t)$$

lying in the region bounded by the ray $x = \xi x_0$, $y = \xi y_0$ $(0 \leq \xi \leq t_1)$ to the point (x_1, y_1) and the half-trajectory $x(x_1, y_1, t)$, $y(x_1, y_1, t)$ for $0 \leq t < \infty$. This half-trajectory of system (25.35) cannot leave this region if $x^2 + y^2 \geq x_0^2 + y_0^2 = r_0^2$. If it did so, it would have to intersect the trajectory $x(x_1, y_1, t)$, $y(x_1, y_1, t)$ for the first time for some $t > 0$. But this is clearly impossible, since each trajectory of system (25.26) intersects the trajectories of system (25.35) in the same direction. On the other hand, the half-trajectory above in question cannot lie entirely in this region for $x^2 + y^2 \geq r_0^2$, since this region contains no singular point or isolated trajectory of system (25.35). The fact that singular points of system (25.35) are absent is shown by relation (25.27). If there were an isolated trajectory in the region $x^2 + y^2 \geqq r_0^2 > 0$, then, since every trajectory of the system (25.26) would cut a trajectory of the system (25.35) in the same direction, the Poincaré index of the region bounded by this trajectory would be different from 0 [138, pp. 133–39]. Therefore the interior of such an isolated trajectory would have to contain a singular point of system (25.26), which is impossible [138]. The orthogonal trajectory (25.34a) must therefore contain a point x_2, y_2 such that the relation $x_2^2 + y_2^2 = r_0^2$ holds. (See Fig. 5.)

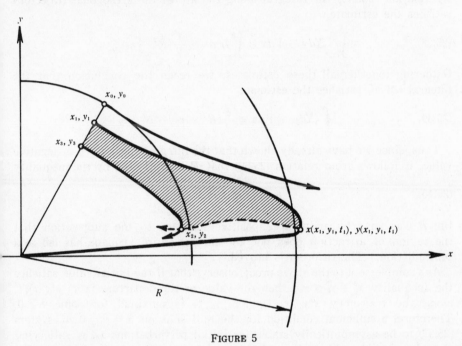

FIGURE 5

The shaded region in the figure is bounded by a segment of the trajectory $x(x_1, y_1, t)$, $y(x_1, y_1, t)$ for $0 \leqq t \leqq t_1$; a segment of the orthogonal trajectory (25.34a); a segment of the trajectory $x(x_2, y_2, t)$, $y(x_2, y_2, t)$ of system (25.26) for $t_3 \leqq t < 0$; and a segment of the line $x = \xi x_0$, $y = \xi y_0$. Here t_3 is defined by the relations $x_3 = x(x_2, y_2, t_3)$, $y_3 = y(x_2, y_2, t_3)$. Let the contour of this region be called Γ. The value of the integral

(25.36)
$$\oint_\Gamma X dy - Y dx$$

is found by Green's Formula to be

$$\oint_\Gamma X dy - Y dx = \iint_{(D)} \left(\frac{\partial X}{\partial x} + \frac{\partial Y}{\partial y} \right) dx dy ,$$

where D is the simply connected region bounded by Γ. Condition (25.28) shows that this value is negative. On the other hand, the value of the integral (25.36) can be computed directly. Along a trajectory of system (25.26), the value of the integral is 0. For the straight-line segment $(x_1, y_1) - (x_3, y_3)$ the Schwartz inequality yields the estimate

$$(25.37) \qquad \left| \int (Y dx - X dy) \right| < r_0 \max [\sqrt{X^2 + Y^2}] = N,$$

the maximum being computed over values of x, y for which $x^2 + y^2 = r_0^2$. By relation (25.27), the integral along the arc of the orthogonal trajectory satisfies the estimate

$$(25.38) \qquad \int_{x^*y^*} X dy - Y dx \geq \int_{r_0}^{R} lr\, dr > \frac{l}{2}(R^2 - r_0^2).$$

Gathering together all these estimates, we reach the conclusion that the integral (25.36) satisfies the estimate

$$(25.39) \qquad \oint_r X dy - Y dx \geq \frac{l}{2}(R^2 - r_0^2) - N.$$

Thus, since we have already shown that the integral must have a negative value, it follows from relation (25.39) that R is restricted by the inequality

$$(25.40) \qquad R \leq \sqrt{r_0^2 + \frac{2N}{l}}.$$

But R was an arbitrarily large positive number, so the supposition that the region of attraction does not contain the entire xy-plane has led to a contradiction. The theorem is proved.

As a complement to the above proof, observe that if the initial value satisfies the inequality $x_0^2 + y_0^2 \leq r_0^2$, then the value of the departure $[x^2(t) + y^2(t)]^{1/2}$ along the trajectory $x(x_0, y_0, t)$, $y(x_0, y_0, t)$ is maximal for some $t \geq 0$. Therefore, a sufficient condition for the null solution $x = y = 0$ of system (25.27) to be asymptotically stable for initial perturbations x_0, y_0 satisfying $x_0^2 + y_0^2 \leq r_0^2$ is that conditions (25.27) and (25.28) of the theorem be satisfied in the region $x^2 + y^2 \leq R^2$, where R satisfies inequality (25.40).

Now we return to system (25.21). Note that the first condition of (25.23) is simply relation (25.28). We shall show that condition (25.27) is a consequence of the second relation of (25.23) if we suppose in addition that $h_1 = f_1(y)/y$ and $h_2 = f_2(x)/x$ are bounded. Consider the system of equations

$$\xi = ax + h_1 y, \qquad \eta = h_2 x + by.$$

Solving this system for x and y, we obtain

$$x = \frac{\xi b - h_2 \eta}{ab - h_1 h_2}, \qquad y = \frac{a\eta - \xi h_2}{ab - h_1 h_2}.$$

Thus relation (25.23) shows that

$$(x^2 + y^2)r^2 = \alpha_1 \xi_1^2 + \alpha_2 \xi \eta + \alpha_3 \eta^2 \leq P(\xi^2 + \eta^2) = P(X^2 + Y^2).$$

This establishes our assertion.

COROLLARY 25.2. *The null solution* $x = y = 0$ *of system* (25.21) *is asymptotically stable for arbitrary initial perturbations if the following conditions are satisfied*:

(i) $$a + b < 0 , \qquad ab - \frac{f_1(x)}{x} \frac{f_2(y)}{y} > 0 ;$$

(ii) *the functions* $h_1 = f_1(x)/x$, $h_2 = f_2(y)/y$ *are bounded for all* $x, y \neq 0$;
(iii) *the limits* $h_1(0) = f_1'(0)$, $h_2(0) = f_2'(0)$ *exist.*[*]

Mangasarian [126a] has proved a theorem concerning global asymptotic stability of system (1.3) under very weak hypotheses. He supposes only that the $X_i(x, t)$ are continuous at $x = 0, 0 \leq t < \infty$, that every solution germ can be indefinitely prolonged, and that for all t the scalar function

$$x_1 X_1(x, t) + \cdots + x_n X_n(x, t)$$

is everywhere a strictly concave function of the variables $x_1, \cdots, x_n$, and that this same scalar function has for $x \neq 0$ a negative upper limit when $t \to \infty$. To establish this result, he uses the Lyapunov function $x^* I x = x_1 x_1 + \cdots + x_n x_n$. Corresponding results obtain when the Lyapunov function $x^* A x$ is used in the same way. Here A is any positive-definite matrix.

Lur'e [113], Malkin [120], Letov [108]–[110], Spasskiĭ [174], and many others have stated the stability problem for regulators (servomechanisms) in a somewhat different formulation from that of Aĭzerman. We shall next give an account of their work and some applications of their method of constructing a Lyapunov function v in the form of the sum of a positive-definite quadratic form and an integral $\int_0^\sigma f(\sigma) \, d\sigma$.

To illustrate the method, we shall construct a Lyapunov function in a form described in the work of Malkin and Lur'e. The construction of the function v in Lur'e's form has been discussed from a very general point of view by Yakubovič [189], [190], who has given some invariant criteria for the possibility of constructing such a Lyapunov function, and has also given a uniform method for constructing it. We consider the system

$$(25.41) \qquad \frac{d\eta_i}{dt} = \sum_{j=1}^{n} a_{ij}\eta_j + h_i f(\sigma) , \quad \sigma = \beta_1\eta_1 + \cdots + \beta_n\eta_n \qquad (i = 1, \cdots, n) ,$$

where a_{ij}, h_i, and β_i are constants, and $f(\sigma)$ is a continuous function[**] satisfying the relation

$$(25.42) \qquad \sigma f(\sigma) > 0 \qquad (\sigma \neq 0) .$$

Lur'e's formulation [113] was designed to determine conditions under which the solution $\eta_1 = \cdots = \eta_n = 0$ is stable for arbitrary initial perturbations

---

[*] The conditions of corollary 25.2 are somewhat weaker than conditions (25.23), the conditions initially introduced on p. 113 as being sufficient for the stability of system (25.21).

[**] In singular cases, the function $f(\sigma)$ can also have finite discontinuities (see, for example, [109]). Our arguments include this case, if we use the remark of section 17, pp. 77-79.

and for arbitrary choice of the function $f(\sigma)$ subject to condition (25.42). To solve this problem, Lur'e made use of a special transformation of the equation and gave a method for constructing a Lyapunov function. Here we give a method expounded by Malkin [120], following Lur'e's ideas.

We define r by

$$(25.43) \qquad r = -\sum_{j=1}^{n} \beta_j h_j \; ;$$

let us further suppose that $r > 0$, and that the roots of the determinantal equation

$$(25.44) \qquad \det\,[a_{ik} - \rho\delta_{ik}]_1^n = 0$$

have negative real part.* We construct a negative-definite quadratic form

$$(25.45) \quad w(\eta_1, \cdots, \eta_n) = -\frac{1}{2}\sum_{i,j=1}^{n} A_{ij}\eta_i\eta_j$$

$$= -\alpha_{11}\eta_1^2 \mp (\alpha_{21}\eta_1 + \alpha_{22}\eta_2)^2 \mp \cdots \mp (\alpha_{n1}\eta_1 + \cdots + \alpha_{nn}\eta_n)^2$$

and choose the positive-definite form

$$(25.46) \qquad F(\eta_1, \cdots, \eta_n) = \sum_{i,j=1}^{n} B_{ij}x_ix_j$$

to satisfy the relation

$$(25.47) \qquad \sum_{i=1}^{n}\frac{\partial F}{\partial \eta_i}(a_{i1}\eta_1 + \cdots + a_{in}\eta_n) = w(\eta_1, \cdots, \eta_n) \; .$$

We consider the function

$$(25.48) \qquad v(\eta_1, \cdots, \eta_n) = \int_0^\sigma f(\sigma)\,d\sigma + F(\eta_1, \cdots, \eta_n) \; .$$

The calculation of the value of dv/dt along a trajectory of system (25.48) is as follows:

$$(25.49) \qquad \frac{dv}{dt} = w(\eta_1, \cdots, \eta_n) + rf^2(\sigma) - f(\sigma)\sum_{i=1}^{n}\left(h_i\frac{\partial F}{\partial \eta_i} + \beta_i\sum_{j=1}^{n}a_{ij}\eta_j\right).$$

The conditions of theorem 5.2 will be satisfied throughout the entire space $-\infty < \eta_i < \infty$, $i = 1, \cdots, n$, if the quadratic form in the arguments $\eta_1, \cdots, \eta_n$, $f(\sigma)$ in the right member of (25.49) is negative-definite. A sufficient condition for this is Sylvester's condition, which takes the form

$$(25.50) \qquad D_1 = A_{11} > 0, \quad D_2 = \det\begin{bmatrix} A_{11} & A_{12} \\ A_{21} & A_{22} \end{bmatrix} > 0, \quad \cdots,$$

$$D_n = \det\begin{bmatrix} A_{11} & A_{12} & \cdots & A_{1n} \\ \cdots\cdots\cdots\cdots\cdots \\ A_{n1} & A_{n2} & \cdots & A_{nn} \end{bmatrix} > 0 \; ,$$

and

---

* Malkin [120] also considered the cases in which equation (25.44) has one or two zero roots.

$$(25.51) \qquad D_{n+1} = \det \begin{bmatrix} A_{11} \cdots A_{1n} \ P_1 \\ A_{21} \cdots A_{2n} \ P_2 \\ \cdots\cdots\cdots\cdots \\ A_{n1} \cdots A_{nn} \ P_n \\ P_1 \ \cdots \ P_n \ \ r \end{bmatrix} > 0 \ ,$$

where

$$(25.52) \qquad P_i = -\sum_{j=1}^{n}(\beta_j a_{ji} + h_j B_{ji}) \qquad (i = 1, \cdots, n) \ .$$

Inequalities (25.50) are automatically satisfied, since w was taken to be a negative-definite form. Thus the null solution $\eta_1 = \cdots = \eta_n = 0$ of system (25.41) will be asymptotically stable for all perturbations and for arbitrary $f(\sigma)$ satisfying condition (25.42), provided that condition (25.51) is satisfied.

The physical system is describable by a set of parameters α_{jk}, and the generality embodied by these parameters is reflected in the appearance of the parameters B_{ij} in the relation (25.52) which describes the regulatory system.

At this point, we mention certain results obtained by Yakubovič. A matrix formulation of his results is convenient. Capital Roman letters denote matrices; boldface lower-case letters denote vectors; Greek letters denote scalar quantities. In addition, the symbol t denotes the time, n the order of the vectors and matrices, V a Lyapunov function, and $dV/dt = -W$ its total derivative. The symbol * denotes the operation of transposition. The symbol (a, b) denotes the scalar product, and the symbol (Hx, x) denotes a quadratic form in the arguments $x_1, \cdots, x_n$: $(Hx, x) = x^*Hx$.

Using this notation, we can write the system of differential equations under consideration in the form

$$\frac{dx}{dt} = Ax + a\varphi(\sigma) \ , \qquad \frac{dz}{dt} = (b, x) - \rho\varphi(\sigma) \ ,$$

where $\rho > 0$. The function $\varphi(\sigma)$, the characteristic of the servomotor, is supposed to be continuous and to satisfy the conditions

$$\varphi(0) = 0 \ , \qquad \sigma\varphi(\sigma) > 0 \text{ if } \sigma \neq 0 \ .$$

We also suppose, as we did above, that all the roots of the characteristic equation of the system $\det[A - \lambda B] = 0$ have negative real part. This requirement corresponds to the formulation of Lur'e's problem, in which we assumed that the solution $x = 0$, $\sigma = 0$ was globally stable, i.e, stable, for arbitrary initial perturbations.

Now we shall repeat the construction of the Lyapunov function V in matrix notation. This formulation is extremely compact, and its form is invariant. The formula for the Lyapunov function V is

$$V = (Hx, x) + \int_0^\sigma \varphi(\sigma)\, d\sigma \ , \qquad H^* = H \ .$$

The total derivative dV/dt of this function has the form

$$\frac{dV}{dt} = -W = -(x^*Gx + 2g^*x + \gamma) ,$$

where $-G = A^*H + HA$, $-g = -(Ha + \frac{1}{2}b)\varphi$, $\gamma = \rho\varphi^2$. We remark that whenever we are discussing global stability (that is, stability for arbitrarily great initial perturbations), the conditions $V > 0$, $W > 0$ for $|\sigma| + ||x|| \neq 0$ guarantee the result without the additional requirement that the function V admit an infinitely great lower bound. (See theorem 5.2.) This assertion is proved by Yakubovič [189], [190]. Thus, to obtain a condition for stability it is desirable to discover under what circumstances the inequalities $V > 0$, $W > 0$ are actually satisfied for $||x|| + |\sigma| \neq 0$. Moreover, by Lyapunov's theorem [114], it is easy to see that the inequality $V > 0$ follows from the inequality $W > 0$. A necessary and sufficient condition for the inequality $W > 0$ to be satisfied is that the relation

$$(25.53) \qquad \gamma G - gg^* = \tfrac{1}{4}\rho\varphi^2 G > 0$$

hold. Following Yakubovič, we write $G > 0$ if G is positive-definite, $(Gz, z) > 0$ for $z \neq 0$. We now write inequality (25.53) explicitly, and we obtain the following quadratic inequality for the matrix H:

$$(25.54) \qquad -\rho(A^*H + HA) - (Ha + \tfrac{1}{2}b)(Ha + \tfrac{1}{2}b)^* \equiv \tfrac{1}{2}\rho G > 0 .$$

Thus a sufficient condition for global stability is that equation (25.54) be valid for some matrix $G > 0$ that can arise; for this it is sufficient that there should be a symmetric matrix H for which the matrix G of (25.54) is positive-definite. A convenient formulation of this condition can be given by inventing a linear operator $Y = R(X)$ defined by the formula $A^*Y + YA = -X$. We also write $-u = Ha + \frac{1}{2}b$. Thus we obtain the relations

$$(25.55) \qquad \begin{aligned} A^*Y + YA &= -X , \\ -u &= Ha + \tfrac{1}{2}b , \\ \rho H &= R(u, u^*) + \tfrac{1}{2}\rho R(G) , \end{aligned}$$

and

$$(25.56) \qquad R(u, u^*)a + \rho u + \tfrac{1}{2}\rho(b + c) = 0 ,$$

where $c = R(G)a$. If the quadratic vector equation (25.56) has a real solution, then the matrix $H > 0$ can be obtained from equation (25.55), and it will satisfy inequality (25.54). Equation (25.56) can in its turn be written in the form

$$(25.57) \qquad \begin{aligned} A^*U + UA &= -uu^* , &\quad \text{(a)} \\ Ua + \rho u + \tfrac{1}{2}\rho(b + c) &= 0 , &\quad \text{(b)} \end{aligned}$$

where the vector c is defined by the equations

$$(25.58) \qquad \begin{aligned} A^*T + TA &= -G < 0 , &\quad \text{(a)} \\ c &= Ta . &\quad \text{(b)} \end{aligned}$$

Equation (25.56) is the indicial equation for (b) in equation (25.57). If (25.56) has a real solution u for some fortunate choice of the vector c defined by formula (25.58), where $G > 0$ can be an arbitrary positive-definite symmetric matrix, then the trivial solution $x = \sigma = 0$ of the system under consideration is globally stable.

It should be emphasized that all calculations needed to discuss the stability of the system can be carried out in a finite number of steps. The details are as follows. We take $c = 0$ in equations (25.56) and (25.57) and try to solve these equations not only when b is the given vector but also for all choices of b sufficiently close to the given vector. This amounts to taking $G = \varepsilon E$, where $\varepsilon > 0$ is a sufficiently small number. We write out equations (25.56) and (25.57) and obtain a system of algebraic equations that can be solved in real numbers and that give conditions for global asymptotic stability. Furthermore, if we suppose that the matrix A can be reduced to diagonal form, then we obtain a solution of the equations in the form given by Lur'e [113].

26. Singular Cases

In this section we discuss the method used in the preceding section for constructing a Lyapunov function V in a singular case. We consider the system of equations

$$(26.1) \qquad \frac{dx_i}{dt} = X_i(x_1, \cdots, x_{n+1}) \qquad (i = 1, \cdots, n + 1),$$

$$X_i(0, \cdots, 0) = 0,$$

where the X_i are continuous differentiable functions in some neighborhood of the point $x_1 = \cdots = x_{n+1} = 0$. Lyapunov [114, p. 299] used the term *singular* for the case in which the characteristic equation

$$(26.2) \qquad \det \begin{bmatrix} a_{11} - \lambda \cdots a_{1,n+1} \\ \cdots\cdots\cdots\cdots\cdots \\ a_{n+1,1} \cdots a_{n+1,n+1} - \lambda \end{bmatrix} = 0$$

of the linearized system

$$(26.3) \qquad \frac{dx_i}{dt} = a_{i1}x_1 + \cdots + a_{i,n+1}x_{n+1}$$

$$\left(a_{ij} = \left[\frac{\partial X_i}{\partial x_j} \right]_{x_j = 0} ; \quad i, j = 1, \cdots, n + 1 \right)$$

has roots λ_i with vanishing real part, but has no root λ_j with positive real part.

We know, moreover, that it is impossible to make an assertion about the stability or instability of the null solution $x = 0$ of system (26.1) if we consider only the first-order approximation (26.3) to the trajectories. To investigate the singular case it is therefore necessary to have a criterion that applies essentially to nonlinear systems which undergo large initial perturbations. Simple linearization is not a sufficient tool.

Here we indicate one criterion for asymptotic stability or instability in the singular case, when one of the roots is zero. This criterion is a broader one than the classical criterion of Lyapunov [114, p. 301] and similar criteria of Malkin [118], [124], Kamenkov [76], and others who supposed that the right members of the equations of perturbed motion are holomorphic functions. Indeed, this criterion can be used when $X_i(x_1, \cdots, x_{n+1})$ are not necessarily analytic functions.

We now return to system (26.1). Suppose that the characteristic equation (26.2) of the linearized system has a single zero root and n roots with negative real part. Lyapunov [114] solved the problem of the stability of the null solution $x_1 = \cdots = x_{n+1} = 0$ when the functions X_i are holomorphic. Vedrov [180] generalized Lyapunov's results to the case in which the functions X_i are supposed only to be differentiable in some neighborhood of the origin $x = 0$. The following theorem is still more general.

THEOREM 26.1. *Suppose that all the roots of the determinantal equation*

(26.4)
$$\det \left[\frac{\partial X_i}{\partial x_j} - \lambda \delta_{ij} \right]_1^{n+1} = 0$$

which depends on x have negative real part in some neighborhood of the origin, $\| x \| \neq 0$. Suppose further that equation (26.2) has a single zero root. Then the null solution $x = 0$ of the system (26.1) is asymptotically stable. Similarly, if in every small neighborhood of the origin equation (26.4) has a root with positive real part, then the null solution $x = 0$ is unstable.

This theorem was proved by the author in [86]. A corresponding theorem has been established recently by Vinograd [182] and the author [82] for a system of two equations, but encompassing more general criteria.

Without giving the details of the proof of theorem 26.1, we indicate the kernel of the proof, the method of constructing the appropriate Lyapunov function V. The main idea was explained in the beginning of the chapter: we insert parameters and determine what values they should have so that the corresponding function is a Lyapunov function.

The author showed in [86] that equation (26.1) can be brought into the following form by a coordinate transformation:

(26.5)
$$\frac{dx}{dt} = \varphi(x) + p_1(x, z_1, \cdots, z_n)z_1 + \cdots + p_n(x, z_1, \cdots, z_n)z_n \,,$$
$$\frac{dz_i}{dt} = [p_{i1} + q_{i1}(x, z_1, \cdots, z_n)]z_1 + \cdots$$
$$+ [p_{in} + q_{in}(x, z_1, \cdots, z_n)]z_n + q_i(x, z_1, \cdots, z_n)\varphi(x)$$
$$(p_{ij} = \text{const.}; \ i = 1, \cdots, n).$$

Here the functions q_{ij}, p_i, q_j approach zero as the corresponding arguments approach the origin. Moreover, the roots of the equation

(26.6)
$$\det [p_{ij} - \lambda \delta_{ij}]_1^n = 0$$

have negative real part. If the roots of equation (26.4) have negative real part, the inequality

(26.7)
$$x \, \varphi(x) < 0$$

will hold for $| x | < \delta, x \neq 0$. On the other hand, if equation (26.4) has a root with positive real part, the contrary relation

(26.8) $$x\,\varphi(x) > 0$$

will hold for $|x| < \delta$, $x \neq 0$, where $\delta > 0$ is a sufficiently small constant.

By Lyapunov's theorem [114, p. 276], there exists a positive-definite quadratic form in the arguments $z_1, \cdots, z_n$:

$$w(z) = \sum_{i,j=1}^{n} b_{ij} z_i z_j$$

such that the relation

(26.9) $$\sum_{j=1}^{n} \frac{\partial w}{\partial z_j} \left(\sum_{i=1}^{n} p_{ji} z_i \right) = -\sum_{i=1}^{n} z_i^2$$

is satisfied.

Consider the function

(26.10) $$v(x, z) = -\int_0^x \varphi(\xi)\, d\xi + w(z) .$$

The total derivative dv/dt along a trajectory of (26.1) [that is, along a trajectory of (26.5)] can be written in the form

$$\frac{dv}{dt} = \varphi^2(x) \sum_{i=1}^{n} p_i z_i \varphi(x) + \sum_{i=1}^{n} \frac{\partial w}{\partial z_i} \left(\sum_{j=1}^{n} p_{ij} z_j \right)$$

$$+ \sum_{i=1}^{n} \frac{\partial w}{\partial z_i} \left(\sum_{j=1}^{n} q_{ij} z_j \right) + \sum_{i=1}^{n} \frac{\partial w}{\partial z_i} q_i \varphi(x) .$$

In a sufficiently small neighborhood of the origin $x = 0$, the functions p_i, q_i, q_{ij} are small. Thus (26.9) shows that the derivative dv/dt is a negative-definite function. If (26.7) holds, the function $v(x, z)$ is itself a positive-definite function. Thus the conditions of Lyapunov's theorem on asymptotic stability [114] are satisfied. If (26.8) holds, the function $v(x, z)$ satisfies the conditions of Četaev's theorem on instability [36, p. 34]. Thus the theorem is proved.

We mention one consequence of theorem 26.1. Consider the equation

(26.11) $$\frac{d^n x}{dt^n} = X\left(x, \frac{dx}{dt}, \cdots, \frac{d^{n-1}x}{dt^{n-1}} \right) .$$

Suppose the characteristic equation of the first-order approximation to this equation has one zero root and n roots with negative real part. Set $y_k = d^k x/dt^k$. A sufficient condition for the solution $x = y_1 = \cdots = y_n = 0$ of equation (26.11) to be asymptotically stable is that the inequality

(26.12) $$\frac{\partial X}{\partial x} < 0 \quad \text{for } x \neq 0$$

should hold in some neighborhood of the point $x = y_1 = \cdots = y_n = 0$. This is really a special case of theorem 26.1. We verify this by checking the Routh-Hurwitz inequalities [36] for the coefficients of equation (26.4) in the special case (26.11).

6

General Theorems Concerning Lyapunov's Second Method for Equations with Time Delay

27. Preliminary Remarks

In this chapter and the next, we shall consider the problem of stability for differential equations with delay in the time t. These equations are important for their potential applications to the theory of automatic regulators or servomechanisms. First, we define the general form of an equation with time delay, following Myškis [136], who called such equations "equations with a past history." We consider the equations

$$(27.1) \qquad \frac{dx_i}{dt} = X_i(x_1(t + \vartheta), \cdots, x_n(t + \vartheta), t) \qquad (i = 1, \cdots, n),$$

where the right-hand members $X_i(x_1(\vartheta), \cdots, x_n(\vartheta), t)$ are functionals defined for piecewise-continuous functions $x_i(\vartheta)$ of the argument ϑ, which is restricted to the interval $-h \le \vartheta \le 0$, where h is a positive constant ($h > 0$).

In equation (27.1) the rate of change of the variable $x_i(t)$ is determined by the values of these variables (that is, by their behavior at some previous moments of time, i.e., for $t + \vartheta$, $-h \le \vartheta \le 0$). We shall always assume (unless the contrary is stated) that a function of the argument ϑ is considered for the values $-h \le \vartheta \le 0$. To determine the derivative dx_i/dt at a given moment of time, it is necessary to know the values of the functions $x_i(t + \vartheta)$ $(i = 1, \cdots, n)$ which describe the behavior of the system at those previous moments of time that are involved in the functional X_i. The numerical value of this functional then gives the value of the derivative dx_i/dt at the moment t. A special case of equation (27.1) is equation (27.2), involving time delay:

$$(27.2) \qquad \frac{dx_i}{dt} = X_i\left(x_1(t - h(t)), \cdots, x_n(t - h(t)), t\right) \qquad (i = 1, \cdots, n),$$

where the derivative dx_i/dt is determined by the value of the function $x_i(t)$ at the preceding moment of time $t - h(t)$. (Here X_i is an ordinary function of the numerical vector-argument $x_1, \cdots, x_n$.) It is also acceptable that each function x_j appear in X_i with its individual delay $h_j(t)$ or even with several time delays.

Another common form of equation (27.1) is

(27.3) $$\frac{dx_i}{dt} = X_i(\theta_1, \cdots, \theta_k, t) \qquad (i = 1, \cdots, n),$$

where X_i is a function of the numerical vector $\{\theta_1, \cdots, \theta_k\}$, and $\theta_1, \cdots, \theta_k$ are functions of the form

(27.4) $$\theta_j = \int_{t-h}^{t} f_j(x_j(\vartheta)) \, d\phi_j(\vartheta),$$

and so on. For $h = 0$, (27.1) reduces to an ordinary differential equation.

In the following sections we shall often be concerned with a function space $\{x_i(\vartheta)\}$ $(-h \leq \vartheta \leq 0, i = 1, \cdots, n)$. For this we need certain notations. We shall not specify ahead of time which one of the standard spaces $(C, L, L_2,$ etc.) we are considering, and for purposes of illustration shall select the one that is convenient for the problems we are considering.

Let us use the following notation:

(27.5) $$\|x\|^{(h)} = \sup \left(|x_i(\vartheta)| \quad \text{for} \quad -h \leq \vartheta \leq 0, \quad i = 1, \cdots, n \right),$$

(27.6) $$\|x\|_2^{(h)} = \left[\int_{-h}^{0} \left(\sum_{i=1}^{n} x_i^2(\vartheta) \right) d\vartheta \right]^{1/2}.$$

We shall reserve the symbols $\|x\|$ and $\|x\|_2$ without the superior index (h) to denote the norms of the numerical-valued vector $\{x_i\}$, which we have already defined (see p. 2). As a rule, we shall be concerned with metrics defined by the norm given in (27.5). We assume further that the functional X_i is defined for

(27.7) $$\|x\|^{(h)} < H, \qquad t \geq 0,$$

where H is a fixed constant (or perhaps $H = \infty$), and that X_i satisfies the following conditions:[*]

(1) The functional X_i is piecewise-continuous in region (27.7).

(2) There is a sequence of numbers t_k $(k = 0, 1, 2, \cdots)$ such that in every region

(27.8) $$\|x(\vartheta)\|^{(h)} < H, \qquad t_k \leq t < t_{k+1}$$

X_i is continuous in t.

(3) The definition of X_i can be continuously extended (that is, the extended definition of X_i preserves the continuity property) into every region $\|x\|^{(h)} < H$, $t_k \leq t \leq t_{k+1}$ in such a way that for every $t^* \in [t_k, t_{k+1}]$ and for every continuous function $x_i^*(\vartheta), i = 1, \cdots, n$, for every positive number $\varepsilon > 0$ we can find a positive number $\delta > 0$ such that the inequalities

$$|X_i(x_1^*(\vartheta), \cdots, x_n^*(\vartheta), t^*) - X_i(x_1^*(\vartheta), \cdots, x_n^*(\vartheta), t)| < \varepsilon \quad (i = 1, \cdots, n)$$

hold whenever $|t^* - t| < \delta$ and $t \in [t_k, t_{k+1}]$. Because the functional X_i may have discontinuities at particular moments of time t_k, and for certain other reasons, which will become clear on pp. 129 and 144, we require only

---

* It will usually be assumed that $X_i(x(\vartheta), t) = 0$ on the entire arc $x_j(\vartheta) \equiv 0; i = 1, \cdots, n;$ $j = 1, \cdots, n; -h \leq \vartheta \leq 0.$

that dx_i/dt in equation (27.1) exist as the right-hand derivative of x_i with respect to the variable t.

(4) The functional X_i satisfies a Lipschitz condition with respect to $x_j(\vartheta)$, uniformly with respect to t, that is,

$$(27.9) \quad | X_i(x_1''(\vartheta), \cdots, x_n''(\vartheta), t) - X_i(x_1'(\vartheta), \cdots, x_n'(\vartheta), t) |$$
$$< L \, || \, x'' - x' \, ||^{(h)} \quad (L = \text{const.}; \; i = 1, \cdots, n) \; .$$

Let us now see when the system (27.1) of equations with time delay admits a solution. Let there be given a certain moment of time $t = t_0 \geq 0$, and the piecewise-continuous functions*

$$x_0(\vartheta_0) \equiv \{x_{i0}(\vartheta_0)\} \qquad (i = 1, \cdots, n; \; -h \leq \vartheta_0 \leq 0) \; ;$$

we call the functions $x(t) \equiv \{x_i(t)\}$ a solution if they are continuous for $t \geq t_0$ and satisfy the condition

$$(27.10) \quad \lim_{\Delta t \to +0} \left(\frac{x_i(t + \Delta t) - x_i(t)}{\Delta t} \right) = X_i(x_1(t + \vartheta), \cdots, x_n(t + \vartheta), t) \; ,$$

provided that for $t + \vartheta \leq t_0$ we have $x_i(t + \vartheta) = x_{i0}(\vartheta_0)$, $\vartheta_0 = t + \vartheta - t_0$.

To emphasize the dependence of $x(t)$ on the initial curve and the initial value of t, we shall denote $x(t)$ by $x(x_0(\vartheta_0), t_0, t)$.

Under the conditions imposed on the functional X_i, the existence of a solution (in the sense of the above definition) can be proved for every piecewise-continuous initial curve $x_0(\vartheta_0)$ for $t \geq t_0$. Moreover, just as for ordinary differential equations, the solution can be continued to all values of $t \geq t_0$ for which the curve $x(x_0(\vartheta_0), t_0, t)$ lies in region (27.7).

We cannot present a proof of this assertion here, since it depends on the general method of successive approximations. (Proofs of theorems of this kind are contained in general works, for instance [39] and [138].) Here we reduce the proof to an auxiliary inequality, which we shall often use later.

LEMMA 27.1. *Let $t = t_0 \geq 0$ be a fixed, positive value of the time t and let the initial curves $\{x_{i0}(\vartheta_0)\}$, $\{x_{i0}^*(\vartheta_0)\}$ satisfy the conditions*

$$|| \, x_0(\vartheta_0) \, ||^{(h)} < H, \qquad || \, x_0^*(\vartheta_0) \, ||^{(h)} < H \; .$$

For all values of $t \geq t_0$ for which the segment of the trajectory $x(x_0(\vartheta_0), t_0, t)$, $x(x_0^*(\vartheta_0), t_0, t)$ lies entirely in the region (27.7), the following inequality holds:

$$(27.11) \quad || \, x(x_0(\vartheta_0), t_0, t) - x(x_0^*(\vartheta_0), t_0, t) \, || \leq || \, x_0(\vartheta_0) - x_0^*(\vartheta_0) \, ||^{(h)} \exp L(t - t_0) \; ,$$

and therefore

$$(27.12) \quad || \, x(x_0(\vartheta_0), t_0, t + \vartheta) - x(x_0^*(\vartheta_0), t_0, t + \vartheta) \, ||^{(h)}$$
$$\leq || \, x_0(\vartheta_0) - x_0^*(\vartheta_0) \, ||^{(h)} \exp L(t - t_0) \; .$$

---

* By a *piecewise-continuous curve* we mean a curve in which the discontinuities are of simple type: the function $x_i(\vartheta)$ may have at most a finite number of points of discontinuity of the first kind (at each such point the limits exist from both right and left). Admitting a piecewise-continuous curve as an initial curve causes no difficulty, and it is necessary to consider such initial curves in the sequel.

Proof. Relation (27.11) is valid for $t = t_0$. Let us suppose on the contrary that the lemma is false, and let $t = t_1$ be a moment of time so defined that in the interval $[t_0, t_1]$ the inequality (27.11) is valid, but for $t > t_1$ it is invalid. That is, for every $\varepsilon > 0$ there is a value t_ε such that whenever

$$(27.13) \qquad || x(x_0(\vartheta_0), t_0, t_\varepsilon) - x(x_0^*(\vartheta_0), t_0, t_\varepsilon) ||$$
$$> || x_0(\vartheta_0) - x_0^*(\vartheta_0) ||^{(h)} \exp L(t_\varepsilon - t_0)$$

we have $|t_\varepsilon - t_1| < \varepsilon$.

From inequality (27.13) we can conclude that for $t > t_1$ the inequality (27.11) is satisfied and we obtain the following:

$$(27.14) \quad \limsup_{\Delta t \to +0} \left(\frac{1}{\Delta t} [|| x(x_0(\vartheta_0), t_0, t_1 + \Delta t) - x(x_0^*(\vartheta_0), t_0, t_1 + \Delta t) ||\right.$$
$$\left. - || x(x_0(\vartheta_0), t_0, t_1) - x(x_0^*(\vartheta_0), t_0, t_1) ||] \right)$$
$$\geq \limsup_{\Delta t \to +0} \left(\frac{1}{\Delta t} [|| x_0(\vartheta_0) - x_0^*(\vartheta_0) ||^{(h)} \{\exp L(t_1 + \Delta t - t_0) - \exp L t_1\}] \right)$$
$$= || x_0(\vartheta_0) - x_0^*(\vartheta_0) ||^{(h)} L \exp L(t_1 - t_0)$$
$$= L || x(x_0(\vartheta_0), t_0, t_1) - x(x_0^*(\vartheta_0), t_0, t_1) || .$$

On the other hand, from equation (27.1), by using condition (27.9), we can reason as follows:

$$(27.15) \quad \sup_{(i=1,\ldots,n)} \left(\left| \frac{dx_i}{dt} - \frac{dx_i^*}{dt} \right| \right)$$
$$= \sup \left[\left| X_i\Big(x(x_0(\vartheta_0), t_0, t_1), t_1 \Big) - X_i\Big(x(x_0^*(\vartheta_0), t_0, t_1), t_1 \Big) \right| \right]$$
$$< L || x(x_0(\vartheta_0), t_0, t_1 + \vartheta) - x(x_0^*(\vartheta_0), t_0, t_1 + \vartheta) ||^{(h)} .$$

Since we are assuming that inequality (27.11) is invalid for $t > t_1$, it follows that

$$\sup \left(|| x(x_0^*(\vartheta_0), t_0, t_1 + \vartheta) - x(x_0(\vartheta_0), t_0, t_1 + \vartheta) || \quad \text{for} \quad -h \leq \vartheta \leq 0 \right)$$
$$= || x(x_0^*(\vartheta_0), t_0, t_1) - x(x_0(\vartheta_0), t_0, t_1) || ,$$

that is,

$$(27.16) \qquad || x(x_0^*(\vartheta_0), t_0, t_1 + \vartheta) - x(x_0(\vartheta_0), t_0, t_1 + \vartheta) ||^{(h)}$$
$$= || x(x_0^*(\vartheta_0), t_0, t_1) - x(x_0(\vartheta_0), t_0, t_1) || .$$

Now by equations (27.15) and (27.16), for every positive γ we can find a positive number $\eta > 0$ such that

$$| [x_i(x_0^*(\vartheta_0), t_0, t_1 + \Delta t) - x_i(x_0(\vartheta_0), t_0, t_1 + \Delta t)]$$
$$- [x_i(x_0^*(\vartheta_0), t_0, t_1) - x_i(x_0(\vartheta_0), t_0, t_1)] |$$
$$< [L || x(x_0^*(\vartheta_0), t_0, t_1) - x(x_0(\vartheta_0), t_0, t_1) || + \gamma] \Delta t$$

for all $i = 1, \cdots, n$, $0 \leq \Delta t \leq \eta$, or

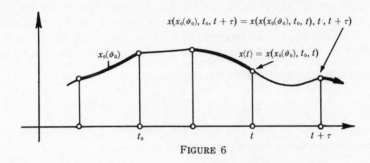

$$x(x_0(\vartheta_0), t_0, t + \tau) = x(x(x_0(\vartheta_0), t_0, t), t, t + \tau)$$

$$x_0(\vartheta_0) \qquad x(t) = x(x_0(\vartheta_0), t_0, t)$$

$$t_0 \qquad t \qquad t + \tau$$

FIGURE 6

(27.17) $\quad \Big| [\,|\, x_i(x_0^*(\vartheta_0), t_0, t_1 + \varDelta t) - x_i(x_0(\vartheta_0), t_0, t_1 + \varDelta t)\,|\,]$

$$- [\,|\, x_i(x_0^*(\vartheta_0), t_0, t_1) - x_i(x_0(\vartheta_0), t_0, t_1)\,|\,] \Big|$$

$$< [L\,||\, x(x_0^*(\vartheta_0), t_0, t_1) - x(x_0(\vartheta_0), t_0, t_1)\,|| + \gamma]\varDelta t$$
$$(i = 1, \cdots, n; \quad 0 \leq \varDelta t \leq \eta)\,.$$

Now let $\varDelta t$ approach $+0$. Then η will approach 0—that is, γ will approach 0—and inequality (27.17) becomes

$$\limsup_{\varDelta t \to +0} \left[\frac{1}{\varDelta t} \Big(||\, x(x_0(\vartheta_0), t_0, t_1 + \varDelta t) - x(x_0^*(\vartheta_0), t_0, t_1 + \varDelta t)\,|| \right.$$

$$\left. -\,||\, x(x_0(\vartheta_0), t_0, t_1) - x(x_0^*(\vartheta_0), t_0, t_1)\,|| \Big) \right]$$

$$< L\,||\, x(x_0^*(\vartheta_0), t_0, t_1) - x(x_0(\vartheta_0), t_0, t_1)\,||\,.$$

But this contradicts inequality (27.14). This contradiction proves the lemma.

In particular, letting $x^*(\vartheta_0) \equiv 0$ in condition (27.11), we obtain the inequality

(27.18) $\qquad ||\, x(x_0(\vartheta_0), t_0, t)\, \leq ||\, x_0(\vartheta_0)\,||^{(h)} \exp L(t - t_0) \quad \text{for} \quad t \geq t_0\,.$

Finally, let us remark that the solution of (27.1) exists, is unique (for $t \geq t_0$), and satisfies the following condition: If $t \geq t_0, \tau > 0$, then

(27.19) $\qquad x(x_0(\vartheta_0), t_0, t + \tau) = x(x_0^*(\vartheta_0), t, t + \tau)\,,$

where

(27.20) $\qquad x_0^*(\vartheta) = x(x_0(\vartheta_0), t_0, t + \vartheta)\,.$

(See Fig. 6.)

For $t < t_0$ and $\tau < 0$ conditions (27.19) and (27.20) may not be satisfied, since the solution may not be continuable to values of the time $t < t_0$.

28. Statement of the Stability Problem for Equations with Time Delay

We define the stability property for equations with time delay.

DEFINITION 28.1. (a) *The solution $x = 0$ of equation* (27.1) *is called stable if for every positive number $\varepsilon > 0$ we can find a positive number $\delta > 0$ such that whenever the inequality*

$$(28.1) \qquad\qquad || x_0(\vartheta_0) ||^{(h)} \leqq \delta$$

is satisfied, the relation

$$(28.2) \qquad\qquad || x(x_0(\vartheta), t_0, t) || < \varepsilon$$

holds for $t \geqq t_0$.

(b) *If, whenever condition* (a) *is satisfied, the conditions*

$$(28.3) \qquad\qquad \lim || x(x_0(\vartheta_0), t_0, t) || = 0 \quad for \quad t \to \infty ,$$

$$(28.4) \qquad || x(x_0(\vartheta_0), t_0, t) || < H_1 \quad for \ all \quad t \geqq t_0 \qquad (H_1 = const.)$$

are satisfied for all initial curves $x_0(\vartheta_0)$ satisfying the inequality

$$(28.5) \qquad\qquad || x_0(\vartheta_0) ||^{(h)} \leqq H_0 ,$$

then the solution $x = 0$ of equation (27.1) *is asymptotically stable and the region* (28.5) *lies in the region of attraction of the unperturbed motion.*[*]

From the formal point of view, definition 28.1 is the most natural extension of Lyapunov's definition of stability to equations with time delay, and this definition is most generally used in work on systems of equations with time delay. (See, for example, [19], [20], [42], [43], [62], [134], [156], [187], and [188].) Nevertheless, it is obvious that there are certain difficulties in this statement of the problem. Indeed, in definition 28.1 it is a requirement that condition (28.1)—or (28.3) or (28.4)—is satisfied for trajectories defined for the very wide class (28.1) or (28.5) of initial curves $x_0(\vartheta_0)$, that is, for every piecewise-continuous initial curve admissible for a given concrete problem. Therefore, a condition guaranteeing the stability for all possible initial curves $x_0(\vartheta_0)$ may be unacceptably narrow in its application to a concrete case.

Consequently, a statement of the stability problem for equations with time delay involves the question: What set of initial curves $x_0(\vartheta_0)$ is sufficiently broad for the problem under consideration? In every case a stability criterion would be considered correct whenever it guaranteed stability for a definite class of initial curves known to include all initial curves arising from possible or admissible perturbations of the given system. The chief practical problem is obviously a statement of the stability problem in a manner or form that is sufficiently wide to include small perturbations of the equation and of the initial curves that are admissible in the structure of the system.

The discussion and exposition that follow will have practical value only

---

[*] In conditions (28.2), (28.3), and (28.4) of definition 28.1, we may write the norms $|| x ||^{(h)}$, which is obviously equivalent to (28.2), (28.3), and (28.4), with changed meaning of the positive numbers ε and δ.

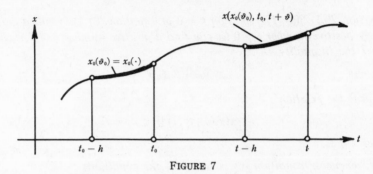

FIGURE 7

for certain general classes of initial curves. However, insofar as there is no fundamental criterion on which to base the choice of these curves, the presentation is carried out in a purely formal sense. Thus we do take the statement of the problem to be that given by definition 28.1.

A sufficient condition for stability according to definition 28.1 is obviously a sufficient condition for a statement of the stability problem that admits only a more restricted class of initial curves. It is also possible that a sufficient condition may turn out to be a necessary condition of stability in such a restricted problem. We remark further that our discussion is restricted to the stability problem. We do not examine criteria for instability.

In stating the stability problem given by definition 28.1, it is convenient to study as an element $x(x_0(\vartheta_0), t_0, t)$ of the trajectory, not the entire vector $\{x_i(x_0(\vartheta_0), t_0, t)\}$, but a vector-segment of this trajectory $\{x_i(x_0(\vartheta_0), t_0, t+\vartheta)\}$ for $-h \leqq \vartheta \leqq 0$ (Fig. 7).

This point of view makes it possible to take advantage of certain results in the theory of semigroups when studying the stability problem for linear systems of equations with time delay. In order to avoid tedious notation we define the following notation for segments of a trajectory:

(1) The vector function $x_i(\vartheta)$, where $i = 1, \cdots, n$ and $-h \leqq \vartheta \leqq 0$, is written $x(\cdot)$.

(2) The initial curve $x_0(\vartheta_0)$ is written $x_0(\cdot)$.

(3) The trajectory $x(x_0(\vartheta_0), t_0, t)$ is written $x(x_0(\cdot), t_0, \cdot)$.

We call attention to the following situation: Suppose an initial piecewise-continuous curve $x_0(\vartheta_0)$ is postulated for time $t = t_0 \geqq 0$. Suppose the trajectory $x(x_0(\vartheta_0), t_0, t)$ lies in region (27.7) for the entire segment $[t_0, \tau]$, where $\tau > t_0 + h$. Then the inequality

$$\left| \frac{dx_i}{dt} \right| = | X_i | < LH = L_1 \qquad (i = 1, \cdots, n)$$

holds for $t = \tau$; that is, for all $t > t_0 + h$ for which the trajectory $x(x_0, t_0, t)$ remains in region (27.7), the functions $x_i(x_0(\vartheta_0), t_0, t)$ satisfy a Lipschitz condition with respect to t and with constant L_1:

(28.6) $| x_i(x_0(\vartheta_0), t_0, t') - x_i(x_0(\vartheta_0), t_0, t'') | < L_1 | t'' - t' |$.

This shows that inequalities (27.11) and (27.12) allow us to use the following line of attack. In a large number of cases, the stability problem can be satisfactorily investigated if we confine ourselves only to initial curves $x_0(\vartheta_0)$ that satisfy a Lipschitz condition. That is, we consider initial curves that satisfy an inequality

(28.7) $\qquad |x_{i0}(\vartheta_0'') - x_{i0}(\vartheta_0')| < L_1 |\vartheta_0'' - \vartheta_0'| \qquad (\vartheta_0'', \vartheta_0' \in [-h, 0])$.

Indeed, let us suppose that for a given value t_0 we have established that the requirements of definition 28.1 are satisfied for trajectories with initial curves which satisfy condition (28.7) and for which, at the initial moment of time t_0' given by $t_0' = t_0 + h$, we have the estimates $\delta(\varepsilon, t_0 + h)$, $H_0(t_0 + h)$ for δ and H_0, respectively, these being the numbers that appear in definition 28.1. Then we can assert that the conditions of definition 28.1 are satisfied for an *arbitrary* piecewise-continuous initial curve, $x_0(\vartheta_0)$, for the initial moment $t = t_0$, with the estimates

$$\delta(\varepsilon, t_0) = \delta(\varepsilon, t_0 + h) e^{-hL} ,$$
$$H_0(t_0) = H_0(t_0 + h) e^{-hL} .$$

Indeed, for conditions (28.2) and (28.3), inequality (27.12) shows that the inequalities

$$\| x(x_0(\vartheta_0), t_0, t_0 + h) \|^{(h)} < \delta(\varepsilon, t_0 + h) ,$$
$$\| x(x_0(\vartheta_0), t_0, t_0 + h) \|^{(h)} < H(t_0 + h)$$

are satisfied. Therefore, the functions $x(x_0(\vartheta_0), t_0, t_0 + \vartheta_0 + k)$ are members of the class of functions (28.7) for $-h \leq \vartheta_0 \leq 0$, for which we supposed the conditions of definition 28.1 to be satisfied. Thus the assertion is established. We shall make use of this remark in the future.

29. Stability of Linear Systems of Equations with Time Delay

We consider the linear system of equations

(29.1) $\qquad \dfrac{dx_i}{dt} = f(x_i(t + \vartheta), \cdots, x_n(t + \vartheta)) \qquad (i = 1, \cdots, n)$

with time delay. Here $f_i(x_1(\vartheta), \cdots, x_n(\vartheta))$ are linear functionals [73, pp. 28–30] of the functions $\{x_i(\vartheta)\}$ ($-h \leq \vartheta \leq 0$). It will be recalled that the term *linear functional* means the following: Let $\{x_i''(\vartheta)\}$ and $\{x_i'(\vartheta)\}$ be a pair of sets of functions, and let c_1 and c_2 be arbitrary constants. Then

(29.2) $\quad c_1 f_i(x_1'(\vartheta), \cdots, x_n'(\vartheta)) + c_2 f_i(x_1''(\vartheta), \cdots, x_n''(\vartheta))$
$$= f_i(c_1 x_1'(\vartheta) + c_2 x_1''(\vartheta), \cdots, c_1 x_n'(\vartheta) + c_2 x_n''(\vartheta)) \qquad (i = 1, \cdots, n) .$$

In accordance with the remark at the end of section 28, we may confine our consideration to the space C of continuous functions $\{x_i(\vartheta)\}$ ($i = 1, \cdots, n$). The norm $\| x(\vartheta) \|^{(h)}$ of the vector function $\{x_i(\vartheta)\}$ is defined by

(29.3) $\quad \| x(\vartheta) \|^{(h)} = \sup (| x_1(\vartheta) |, \cdots, | x_n(\vartheta) |)$ for $i = 1, \cdots, n; -h \leq \vartheta \leq 0)$.

The space of vector functions with norm (29.3) is denoted by the same symbol C.

We assume that the linear functionals f_i are bounded.* This means that the inequalities

(29.4) $|f_i(x_1(\vartheta), \cdots, x_n(\vartheta))| \leqq L_i \|x(\vartheta)\|^{(h)}$ $(i = 1, \cdots, n)$

are satisfied. Here

$$L_i = \sup\left(\frac{f_i(x_1(\vartheta), \cdots, x_n(\vartheta))}{\|x(\vartheta)\|^{(h)}}\right)$$

is the well-known norm of the functional [164]. The theorem of Riesz asserts that the general form of a linear functional over the space of continuous vector functions with norm (29.3) has the form [73], [164]

(29.5) $f_i(x_1(\vartheta), \cdots, x_n(\vartheta)) = \sum_{j=1}^{n} \int_{-h}^{0} x_j(\vartheta)\, d\eta_{ij}(\vartheta)$.

Here the integrals are Stieltjes integrals. In the special case in which the Stieltjes measure satisfies $d\eta_{ij} = 0$ for all but a finite number of values of $\vartheta \neq h_{ij}$, and $d\eta_{ij} = a_{ij}$ for the finite number of values $\vartheta = h_{ij}$, (29.5) has the form

$$f_i(x_1(\vartheta), \cdots, x_n(\vartheta)) = \sum_{j=1}^{n} a_{ij}x_j(t - h_{ij}) ,$$

so that the system of equations (29.1) is the system of linear equations with constant values h_{ij} of the time delay:

(29.6) $\dfrac{dx_i}{dt} = \sum_{i=1}^{n} a_{ij}x_j(t - h_{ij})$ $(i = 1, \cdots, n)$.

The constant values h_{ij} may or may not all be equal to a single constant h.

Linear equations with time delay were considered by Myškis [136]. In this book we consider trajectories of system (29.1) from the point of view of the theory of semigroups of linear operators.

Let $x(x_0(\vartheta_0), t)$ be a trajectory of system (29.1); we denote it by the symbol $x(x_0(\cdot), t, \cdot)$ introduced in the preceding section. For fixed $t \geqq 0$, we consider the element $x(\cdot) = x(x_0(\cdot), t, \cdot)$ to be the image of an element $x_0(\cdot) \in C$ under some transformation $x(\cdot) = T(t)x_0(\cdot)$. This transformation T is a linear one; that is,

(29.7) $T(t)(x'(\cdot)c_1 + x''(\cdot)c_2) = c_1 T(t)x'(\cdot) + c_2 T(t)x''(\cdot)$.

This last condition follows from the linear character of system (29.1). Indeed, on using the definition of the transformation $T(t)$ and the linearity of the functional f_i, we obtain

$$\begin{aligned}
T(t)(x'(\cdot)c_1 + x''(\cdot)c_2) &= x([c_1x'(\cdot) + c_2x''(\cdot)], t, \cdot) \\
&= c_1x(x'(\cdot), t, \cdot) + c_2x(x''(\cdot), t, \cdot) \\
&= c_1 T(t)x'(\cdot) + c_2 T(t)x''(\cdot) ,
\end{aligned}$$

---

* For future convenience we do not include the hypothesis of boundedness as part of the hypothesis of linearity of a functional.

which is a verification of (29.7) based simply on the given equations (29.1). Moreover, equation (27.19) yields

$$(29.8) \qquad T(t + \tau)x_0(\cdot) = x(x_0(\cdot), t + \tau)$$
$$= x(x(x_0(\cdot), t, \cdot), \tau, \cdot)$$
$$= T(\tau)T(t)x_0(\cdot) \quad \text{for} \quad t, \tau > 0 .$$

Thus, for an arbitrary nonnegative value of $t \geq 0$ the transformation of the segment $x_0(\vartheta_0) \equiv x_0(\cdot)$ into the segment $x(x_0(\vartheta_0), t + \vartheta) \equiv x(x_0(\cdot), t, \cdot)$ along a trajectory of the system (29.1) is a linear transformation of the space C into itself. Moreover, the dependence of this transformation on t satisfies condition (29.8). The general name for this last condition is "the semigroup property." Let us review some properties of the transformation $T(t)x(\cdot)$.

The relation

$$(29.9) \qquad || x(\cdot) ||^{(h)} = || T(t)x_0(\cdot) ||^{(h)} \leq || x(\cdot) ||^{(h)} e^{Lt}$$

holds for $t \geq 0$; this follows from the definition of the operator $T(t)$ and from inequality (27.12).

We have $\lim T(t) = J$ as $t \to +0$ in the strong topology [73]; that is, for arbitrary fixed $x_0(\cdot)$, the relation

$$(29.10) \qquad \lim_{t \to +0} || T(t)x_0(\cdot) - x_0(\cdot) ||^{(h)} = 0$$

holds. Here the symbol J denotes the identity operator $Jx(\cdot) = x(\cdot)$. Indeed, by definition of the transformation $T(t)$ and the norm (29.3), we obtain

$$(29.11) \qquad || T(t)x_0(\cdot) - x_0(\cdot) ||^{(h)} = \sup [| x_i(x_0(\vartheta_0), t, \vartheta) - x_{i0}(\vartheta) |]$$
$$(i = 1, \cdots, n; \ -h \leq \vartheta \leq 0) .$$

Let a positive $\varepsilon > 0$ be given. The continuous functions $x_i(x_0(\vartheta_0), t + \vartheta)$ are uniformly continuous on the finite interval of values of time $-h \leq \vartheta \leq \gamma$ where $\gamma > 0$, and therefore, for any given ε, we can find $\delta > 0$ such that

$$(29.12) \qquad | x_i(x_0(\vartheta_0), t + \vartheta_1) - x_i(x_0(\vartheta_0), t + \vartheta_2) | < \varepsilon \quad \text{for} \quad | \vartheta_1 - \vartheta_2 | < \delta .$$

Inequality (29.12) shows that $| x_i(x_0(\vartheta_0), t, \vartheta) - x_{i0}(\vartheta) | < \varepsilon$ whenever $t < \delta$; it also shows, in view of (29.1), that relation (29.10) is valid. This proves the assertion.

Now let us consider the operator $Ax_0(\cdot)$, which is defined by the relation

$$(29.13) \qquad Ax_0(\cdot) = \lim_{\xi \to +0} \left[\frac{1}{\xi} [T(t + \xi)x(\cdot) - T(t)x(\cdot)] \right] .$$

This definition has a formal defect: We may not assume that the limit is necessarily an element of the space C; that is, we may not assume from the start that the operator A, which in general terminology we call an infinitesimal differential operator, is defined for all $x(\cdot)$ of C and defines a transformation of $x_0(\cdot)$ of C into $x(\cdot)$ of C. Moreover, the operator A is not necessarily bounded on C. However, these inconveniences do not prevent our using the operator A. First, we find an explicit expression for A in terms of the functionals f_i appearing in the right member of system (29.1).

According to the definitions of the operators A and $T(t)$ we can write (29.13) in the form

$$Ax_0(\cdot) = \lim_{\xi \to +0}\left[\frac{1}{\xi}[x(x_0(\cdot), \xi, \cdot) - x_0(\cdot)]\right].$$

We could say that the operator A, applied to the functions $x_0(\cdot) \equiv \{x_{i0}(\vartheta)\}$, carries these functions at each point ϑ into their right derivatives (provided that their right derivative exists). Thus, whenever the functions $x_0(\vartheta)$ are differentiable, the operator A is defined and carries each function into its right derivative. But the right derivative of a solution $x(x_0(\vartheta), t)$ for $t = 0$ is equal to the value of f_i, as (29.1) shows. Since A is defined on the differentiable functions $\{x_i(\vartheta)\}$ (although its range includes discontinuous functions), the explicit form of this operator may be found as follows.

The application of the operator A to a differentiable function $\{x_i(\vartheta)\}$ gives a (possibly discontinuous) vector function $\{y_i(\vartheta)\}$, namely

$$y_i(\vartheta) = \frac{dx_i(\vartheta)}{d\vartheta} \quad \text{for} \quad -h \leqq \vartheta < 0,$$

(29.14)
$$y_i(0) = f_i(x_i(\vartheta), \cdots, x_n(\vartheta)),$$

$$y(\cdot) = Ax(\cdot).$$

According to the remark on p. 133, we can restrict ourselves to differentiable initial curves $x_0(\vartheta)$, and therefore we may write equation (29.1) as the equivalent "ordinary differential equation" $dx(t, \cdot)/dt = Ax(t, \cdot)$, defined in the space of differentiable functions $x(\vartheta)$. The stability problem for ordinary differential equations with operator on the right member is investigated by Kreĭn [95]. Here, however, his results are not immediately applicable, since the operator A that he considered is assumed to be bounded. One of the properties of the operator $T(t)$ established above makes it possible to circumvent the difficulty inherent in the possible unboundedness of the operator A. Indeed, we can call on results of the theory of semigroups of linear operators stated by Hille and Phillips [73]. We shall obtain criteria for asymptotic stability for the linear systems (29.1).

THEOREM 29.1. *If the spectrum λ_σ of the operator A (29.14) satisfies the condition*

(29.15) $\text{Re}\,\lambda_\sigma \leqq -\gamma \quad (\gamma = const., \gamma > 0),$

then the solution $x(x_0(\cdot), t, \cdot)$ of the system of equations (29.1) satisfies

(29.16) $\| x(x_0(\cdot), t, \cdot)\| \leqq Be^{-\alpha t}\| x_0(\cdot)\| \quad for \quad t \geqq 0,$

where B, α are positive constants and where, moreover, $\alpha = q\gamma$, with q an arbitrary preassigned number $0 < q < 1$.*

Remark. We recall that the spectrum of the operator A is the set of those complex numbers for which the operator $\lambda J - A$ does not have a

---

* In this section the symbol $\| x(\cdot)\|$ always denotes the norm in the space C, and thus in the norm symbol $\| x\|^{(h)}$, we can omit the index (h).

unique bounded inverse operator in the space C [73, p. 54].   We determine the spectrum of the transformation A from the form (29.14) of the operator A. Let us note the specific form of the operator $\lambda J - A$:

$$y(\cdot) = (\lambda J - A)x(\cdot) ,$$

(29.17)
$$y_i(\vartheta) = - \frac{dx_i}{d\vartheta} + \lambda x_i(\vartheta) \quad \text{for} \quad \vartheta \in [-h, 0) ,$$

$$y_i(0) = \lambda x_i(0) - f_i(x_1(\vartheta), \cdots, x_n(\vartheta)) .$$

By definition, the spectrum $\{\lambda_\sigma\}$ consists of the values λ for which the system of equations (29.17) has no solution for arbitrary $y_i(\vartheta)$, and, when it can be solved, has more than one solution in terms of the given continuous functions $\{y_i(\vartheta)\}$.   The first of equations (29.17) can be solved in the form

(29.18)
$$x_i(\vartheta) = x_i^* e^{\lambda\vartheta} - \int_0^\vartheta e^{\lambda(\vartheta-\tau)} y_i(\tau)\, d\tau \qquad (i = 1, \cdots, n) .$$

Substituting this equation in the second relation of (29.17), we obtain

$$y_i(0) = \lambda x_i^* - f_i\left(x_1^* e^{\lambda\vartheta} - \int_0^\vartheta e^{\lambda(\vartheta-\tau)} y_1(\tau)d\tau, \quad \cdots, \quad x_n^* e^{\lambda\vartheta} - \int_0^\vartheta e^{\lambda(\vartheta-\tau)} y_n(\tau)\, d\tau \right)$$

$$(i = 1, \cdots, n) ,$$

and, using the fact that the functional f_i is linear, and transferring the terms with x_j^* to the left side and the term $y_i(0)$ to the right side, we obtain a system of linear equations for the unknown number x_j^*:

(29.19)
$$\sum_{j=1}^n f_i(0, \cdots, \underset{j}{e^{\lambda\vartheta}}, \cdots, 0)x_j^* - \lambda x_i^*$$

$$= f_i\left(\int_0^\vartheta e^{\lambda(\vartheta-\tau)} y_1(\tau)\, d\tau, \cdots, \int_0^\vartheta e^{\lambda(\vartheta-\tau)} y_n(\tau)\, d\tau \right) - y_i(0) .$$

A unique solution x_j^* of system (29.19) exists if and only if

$$\det \begin{bmatrix} f_{11} - \lambda & \cdots & f_{1n} \\ f_{21} & \cdots & f_{2n} \\ \cdots\cdots\cdots\cdots\cdots \\ f_{n1}\cdots f_{nn} - \lambda \end{bmatrix} \neq 0 ,$$

where

(29.20)
$$f_{ij} = f_i(0, \cdots, \underset{j}{e^{\lambda\vartheta}}, \cdots, 0) .$$

Therefore, the spectrum λ_σ of the operator A is by definition the set of roots of the equation

(29.21)
$$\det \begin{bmatrix} f_{11} - \lambda & \cdots & f_{1n} \\ \cdots\cdots\cdots\cdots\cdots \\ f_{n1} & \cdots f_{nn} - \lambda \end{bmatrix} = 0 .$$

If the functionals f_i have the particular form (29.4), then equation (29.21) takes the form

$$(29.22) \quad \det \begin{bmatrix} \int_{-h}^{0} e^{\lambda\vartheta}\, d\eta_{11}(\vartheta) - \lambda & \cdots & \int_{-h}^{0} e^{\lambda\vartheta}\, d\eta_{1n}(\vartheta) \\ \cdots\cdots\cdots\cdots\cdots\cdots\cdots\cdots\cdots\cdots\cdots \\ \int_{-h}^{0} e^{\lambda\vartheta}\, d\eta_{n1}(\vartheta) & \cdots & \int_{-h}^{0} e^{\lambda\vartheta}\, d\eta_{nn}(\vartheta) - \lambda \end{bmatrix} = 0 \,,$$

and finally, in case the system with time delay has the form (29.5), equations (29.21) can be written

$$(29.23) \quad \det \begin{bmatrix} a_{11}\exp(-\lambda h_{11}) - \lambda & \cdots & a_{1n}\exp(-\lambda h_{1n}) \\ \cdots\cdots\cdots\cdots\cdots\cdots\cdots\cdots\cdots\cdots\cdots \\ a_{n1}\exp(-\lambda h_{n1}) & \cdots & a_{nn}\exp(-\lambda h_{nn}) - \lambda \end{bmatrix} = 0 \,.$$

Criteria for asymptotic stability for systems (29.5) with time delay were obtained by Bellman [19], [20] and Wright [187], [188]. It will be seen that the general approach given above includes all their criteria, since the solution $x = 0$ of the ordinary differential equation $dx/dt = Ax$ is asymptotically stable if the spectrum λ_σ of the operator A satisfies (29.15).

The consideration of trajectories of a system with time delay from the point of view of the theory of semigroups has the additional advantage that all the arguments remain valid in the much more general case when equation (29.1) takes the form

$$(29.24) \qquad dx/dt = X(x(t + \vartheta)) \,,$$

where x is an element of an arbitrary Banach space B, and $X[x(\vartheta)]$ is a linear bounded functional (with values in B) over the space of continuous functions $x(\vartheta)$, $-h \leq \vartheta \leq 0$, with norm $\|x(\vartheta)\|^{(h)} = \sup_{-h \leq \vartheta \leq 0}\|x(\vartheta)\|$. And here equation (29.24) can still be converted into the equivalent ordinary differential equation

$$(29.25) \qquad \frac{dx(t, \cdot)}{dt} = Ax(t, \cdot) \,,$$

where the linear bounded operator A is defined by the relation $y(\cdot) = Ax(\cdot)$, or, in differential form,

$$(29.26) \qquad \begin{aligned} y(\vartheta) &= \frac{dx(\vartheta)}{dt} \quad \text{for} \quad -h \leq \vartheta < 0 \,, \\ y(0) &= X(x(\vartheta)) \,. \end{aligned}$$

The assertion corresponding to theorem 29.1 remains valid for the more general equation (29.24) and takes the following form:

If the spectrum λ_σ of the operator A (29.26) satisfies

$$(29.27) \qquad \operatorname{Re}\lambda_\sigma \leq -\gamma \qquad (\gamma > 0) \,,$$

then the solution $x(x_0(\cdot), t, \cdot)$ of equation (29.24) satisfies

$$(29.28) \qquad \|x(x_0(\cdot), t, \cdot)\| \leq \|x_0(\cdot)\| B e^{-\alpha t} \,,$$

where B, α are positive constants and $\alpha = q\gamma$ for $q \in (0, 1)$.

We do not give a proof of this assertion, since it can be derived by arguments entirely analogous to our proof of theorem 29.1 (see below). Indeed, theorems of functional analysis to which we can refer for the proof of theorem 29.1 remain valid for general Banach spaces (the necessary general form of these theorems is set out and proved in [146], to which we refer the reader).

We note in conclusion that the spectrum λ_σ of the generalization of the operator A [defined by equations (29.14)], i.e., of the operator defined by (29.26), is the set of all the complex numbers λ for which the linear operator

$$\Delta(\lambda)x = X(e^{\lambda\vartheta}x) - \lambda Jx \quad \text{for} \quad x \in B$$

has no bounded inverse (transformation).

PROOF OF THEOREM 29.1.   We consider the equations

$$(29.29) \qquad \frac{dz_i}{dt} = \varphi_i(z_1(t + \vartheta), \cdots, z_n(t + \vartheta)) \qquad (i = 1, \cdots, n) \,,$$

where

$$(29.30) \qquad \varphi_i(z_1(\vartheta), \cdots, z_n(\vartheta)) = f_i(e^{-\gamma\vartheta}z_1(\vartheta), \cdots, e^{-\gamma\vartheta}z_n(\vartheta)) + \gamma z_i(0) \,.$$

Obviously, the solutions $x(x_0(\cdot), t)$ and $z(z_0(\cdot), t)$ of relations (29.1) and (29.28) must satisfy

$$(29.31) \qquad x(x_0(\cdot), t) = z(z_0(\cdot), t)e^{-\gamma t}$$

because of the assumption that the initial curves $x(\cdot)$ and $z(\cdot)$ satisfy

$$(29.32) \qquad x_0(\vartheta) = z_0(\lambda)e^{-\gamma\vartheta} \quad \text{for} \quad -h \leqq \vartheta \leqq 0 \,.$$

We consider the operation $\Delta_0(\lambda)z$, defined on the vector $z = \{z_1, \cdots, z_n\}$ in the following manner:

$$(29.33) \qquad \{\Delta_0(\lambda)z\}_i = \sum_{j=1}^{n} f_i(0, \cdots, e^{(-\gamma)\vartheta}_{j}, \cdots, 0)z_j - (\lambda - \gamma)z_j \,.$$

By assumption, the set of values of λ for which the operator standing on the left member of equation (29.1) has no bounded inverse operator $\Delta^{-1}(\lambda)$ satisfies the inequality $\operatorname{Re}\lambda \leqq -\gamma$, and therefore the set of all values of λ for which the operator (29.33) has no bounded inverse transformation $\Delta_0^{-1}(\lambda)$ satisfies $\operatorname{Re}\lambda \leqq 0$.

Let us denote by A_0 the operator defined as follows:

$$(29.34) \qquad A_0x(\cdot) = y(\cdot) = \begin{cases} y_i(\vartheta) = \dfrac{dx_i(\vartheta)}{d\vartheta} & \text{for} \quad -h \leqq \vartheta < 0 \,, \\[2mm] y_i(0) = \varphi_i(x_1(\vartheta), \cdots, x_n(\vartheta)) \,. \end{cases}$$

The operator A_0 is the infinitesimal differential operator for the semigroup of operators $T_0(t)z_0(\cdot)$, defined on the trajectories of the system of equations (29.29); that is,

$$z(z_0(\cdot), t, \cdot) = T_0(t)z_0(\cdot)$$

and

$$(29.35) \qquad \frac{dz(z_0(\cdot), t, \cdot)}{dt} = A_0 z(z_0(\cdot), t, \cdot) \; .$$

Because of equations (29.29), hypothesis (29.24) shows that the linear functionals φ_i satisfy

$$(29.36) \qquad | \varphi_i(z_1(\vartheta), \cdots, z_n(\vartheta)) | \leq L_0 \, || \, z(\vartheta) \, ||^{(h)} \; ,$$

where $L_0 \leq \max (L_i + \gamma) e^{\gamma h}$.

Consequently, by lemma 27.1, the solution of the system of equations (29.29) satisfies

$$(29.37) \qquad || \, z(z_0(\cdot), t, \cdot) \, || \leq || \, z_0(\cdot) \, || \exp (L_0 t) \quad \text{for } t \geq 0 \; ,$$

and consequently, the semigroup of operators $T_0(t)$ satisfies

$$(29.38) \qquad || \, T_0(t) \, || \leq \exp (L_0 t) \; ,$$

where the norm of the operator $T_0(t)$ stands for

$$|| \, T_0(t) \, || = \sup \frac{|| \, T_0(t) z_0(\cdot) \, ||}{|| \, z_0(\cdot) \, ||} \; .$$

By theorem 11.6.1 of Hille and Phillips [73, p. 349] an operator that satisfies relation (29.38) can be represented in the form

$$(29.39) \qquad T_0(\xi) z(\cdot) = \lim_{\omega \to \infty} \frac{1}{2\pi i} \int_{\gamma_1 - i\omega}^{\gamma_1 + i\omega} e^{\lambda \xi} R(\lambda, A_0) z(\cdot) \, d\lambda \; ,$$

where $\gamma_1 > L_0$, $z(\cdot)$ is a differentiable vector function $\{z_1(\vartheta), \cdots, z_n(\vartheta)\}$, and $R(\lambda, A_0)$ is the resolvent operator of A_0.[*] See also their theorem 12.3.1 [73, p. 360].

The resolvent of an operator A_0 is an operator $R(\lambda, A_0)$ that solves the equation $y(\cdot) = (\lambda J - A_0) x(\cdot)$ in the form $x(\cdot) = R(\lambda, A_0) y(\cdot)$.

We computed the resolvent of the operator A earlier [see formula (29.18)]. The resolvent of the operator A_0 can be obtained in a form similar to (29.18), with the difference that the value of the vector x^* is computed not from the system of equations (29.19), but from the system of equations

$$(29.40) \qquad \sum_{j=1}^{n} \varphi_i(0, \cdots, e^{\lambda \vartheta}, \cdots, 0) x_j^* - \lambda x_j^*$$
$$= y_i(0) - \varphi_i \left(\int_0^\vartheta e^{\lambda(\vartheta - \tau)} y_1(\tau) d\tau, \cdots, \int_0^\vartheta e^{\lambda(\vartheta - \tau)} y_n(\tau) \, d\tau \right) .$$

Since the spectrum of the operator A_0 satisfies the relation $\operatorname{Re} \lambda \leq 0$, it follows that for $\operatorname{Re} \lambda > 0$ the determinant of the system (29.40) is different

---

* The integral in (29.39) is of course an integral in some extended sense since the values of the functions under the integral are not numbers but vector functions. We cannot here go completely into the definition of such integrals (see, for example, Hille and Phillips [73, section 3.7, pp. 76–85]). Here all we need to know is that it is possible to obtain estimates of those integrals in the same way that general estimates of ordinary integrals are obtained.

from zero, and therefore, for every number $\gamma_1 > 0$ we can find a number $P(\gamma_1)$ such that

$$(29.41) \qquad \| R(\lambda, A_0) \| \leq P(\gamma_1) \quad \text{for} \quad \gamma_1 \leq \operatorname{Re} \lambda \leq L_0 + 1 .$$

Moreover, by a result contained in [73], the resolvent $R(\lambda, A_0)$ can be written in the form

$$(29.42) \qquad R(\lambda, A_0) z(\cdot) = \frac{z(\cdot)}{\lambda} + \frac{1}{\lambda^2} A_0 z(\cdot) + \frac{1}{\lambda^2} R(\lambda, A_0) A_0^2 z(\cdot)$$

for all continuous vector functions $z(\cdot) \equiv \{z_1(\vartheta), \cdots, z_n(\vartheta)\}$ for which it is possible, by twice applying the operator A_0, to obtain a function

$$(29.43) \qquad y(\cdot) = A_0 [A_0 z(\cdot)]$$

which is again a continuous vector function. The set of such vector functions $z(\cdot)$ is called the domain of A_0^2 and is denoted by $D[A_0^2]$.

According to the information contained in relations (29.41) and (29.42), we can assert that when operating on vector functions $z(\cdot) \in D[A_0^2]$, the semi-group of operators $T_0(\xi)$ can be represented in the form (29.39) not only for $\gamma_1 > L_0$ but also for arbitrary $\gamma_1 > 0$. If we substitute the solution (29.42) into (29.39), we can obtain the following formula:

$$(29.44) \qquad T_0(\xi) z(\cdot) = \lim_{\omega \to \infty} \frac{1}{2\pi i} \left[\int_{\gamma_1 - i\omega}^{\gamma_1 + i\omega} e^{\lambda \xi} \frac{z(\cdot)}{\lambda} \, d\lambda + \int_{\gamma_1 - i\omega}^{\gamma_1 + i\omega} e^{\lambda \xi} \frac{A_0 z(\cdot)}{\lambda^2} \, d\lambda \right.$$
$$\left. + \int_{\gamma_1 - i\omega}^{\gamma_1 + i\omega} e^{\lambda \xi} \frac{R(\lambda, A_0) A_0^2 z(\cdot)}{\lambda^2} \, d\lambda \right] .$$

From formula (29.44) and inequality (29.41), we have the estimate

$$(29.45) \quad \| T_0(\xi) z(\cdot) \| \leq (\| z(\cdot) \| + P_1 \| A_0 z(\cdot) \| + P_2 \| A_0^2 z(\cdot) \|) \exp (\gamma_1 + \beta) \xi$$

(β is a sufficiently small positive number) for $\xi > 0$, and for an arbitrary vector function $z(\cdot) = z(z_0(\cdot), t, \cdot)$. (Here we use properties of an abstract integral similar to properties of ordinary integrals, as explained in [73].)

We consider the solution $z(z_0(\cdot), t, \cdot)$ of the system of equations (29.29), where $z_0(\cdot)$ is an arbitrary piecewise-continuous function. We note that for $t = 3h$ the function

$$z(\cdot) = z(z_0(\cdot), t, \cdot) \equiv \{z_1(\vartheta), \cdots, z_n(\vartheta)\}$$

lies in the space $D[A_0^2]$, and therefore there are positive constants N_1 and N_2 such that

$$(29.46) \qquad \| A_0 z(z_0(\cdot), 3h, \cdot) \| \leq N_1 \| z_0 \| ,$$

$$(29.47) \qquad \| A_0^2 z(z_0(\cdot), 3h, \cdot) \| \leq N_2 \| z_0 \| .$$

Indeed, the solution $z(z_0(\cdot), t)$ of the system (29.29) is a continuous function of time for $t > 0$. Consequently, we can conclude from (29.29) that the derivatives dz_i/dt are also continuous functions of time. It can be shown [164] that the linear functionals $\varphi_i(z(t + \vartheta))$ satisfy

$$(29.48) \qquad \frac{d\varphi_i}{dt} = \varphi_i\left(\frac{dz(t + \vartheta)}{dt}\right).$$

Therefore, the functions $dz_i/dt = \varphi_i(z(t + \vartheta))$ are continuously differentiable for $t > 2h$; that is, the functions $z(z_0(\cdot), t)$ are twice continuously differentiable functions of the time t for $t > 2h$, and moreover,

$$(29.49) \qquad \frac{dz_i}{dt} = \varphi_i(z(t + \vartheta)),$$

$$(29.50) \qquad \frac{d^2 z_i}{dt^2} = \varphi_i\left(\frac{dz(t + \vartheta)}{dt}\right).$$

Consequently, the functions $z(\vartheta) = z(z_0(\cdot), 3h + \vartheta)$, $-h \leqq \vartheta \leqq 0$, satisfy the following conditions:

(1) The vector function $z(\vartheta)$ is twice continuously differentiable,

$$(2) \qquad \left(\frac{dz_i(\vartheta)}{d\vartheta}\right)_{\vartheta=0} = \varphi_i\left(z_1(\vartheta), \cdots, z_n(\vartheta)\right),$$

$$(3) \qquad \left(\frac{d^2 z_i(\vartheta)}{d\vartheta^2}\right)_{\vartheta=0} = \varphi_i\left(\frac{dz_1(\vartheta)}{d\vartheta}, \cdots, \frac{dz_n(\vartheta)}{d\vartheta}\right).$$

But these three conditions state precisely that the functions $\{z_i(z_0(\cdot), 3h + \vartheta)\}$ are in the domain of the operator A_0^2. It remains to show that there are two numbers N_1 and N_2 satisfying inequalities (29.46) and (29.47). According to inequality (27.12), we have

$$\| z(z_0(\cdot), 3h, \cdot) \| \leqq \| z_0(\cdot) \| \exp(L_0 3h),$$

and from inequality (29.36) and relation (29.50) we conclude that the inequality

$$\left\| \frac{dz(z_0(\cdot), 3h + \vartheta)}{d\vartheta} \right\| \leqq \| z_0(\cdot) \| \exp(L_0 3h) = N_1 \| z_0(\cdot) \|$$

is valid, which proves inequality (29.46). Moreover, from (29.46) and (29.49) we obtain

$$\left\| \frac{d^2 z(z_0(\cdot), 3h +)\vartheta}{d\vartheta^2} \right\| \leqq \| z_0(\cdot) \| N_1 L_0 = N_2 \| z_0(\cdot) \|,$$

which proves inequality (29.47). Therefore our assertion that the function $z(z_0(\cdot), 3h, \cdot)$ is in $D[A_0^2]$ and satisfies conditions (29.46) and (29.47) is completely proved. But from estimates (29.45)–(29.47) we obtain the estimate

$$\| z(z_0(\cdot), t, \cdot) \| = \| T_0(t)z_0(\cdot) \|$$

$$\leqq \Big(\| z(z_0(\cdot), 3h, \cdot) \| + P_1 N_1 \| z(z_0(\cdot), 3h, \cdot) \|$$

$$+ P_2 N_2 \| z(z_0(\cdot), 3h, \cdot) \| \Big) \exp[(\gamma_1 + \beta)(t - 3h)]$$

for $t > 3h$, or, finally,

$$\| z(z_0(\cdot), t, \cdot) \| \leqq Q \exp[(\gamma_1 + \beta)t] \| z_0(\cdot) \| \quad \text{for} \quad t \geqq 0,$$

where Q is a constant and $z_0(\cdot)$ is an arbitrary piecewise-continuous initial curve. From relations (29.31) and (29.32), connecting the solutions of systems (29.1) and (29.39), we can conclude further that the solution of system (29.1) satisfies the estimate

$$(29.51) \quad ||x(x_0(\cdot), t, \cdot)|| \leq Q \exp(\gamma h)||x_0(\cdot)|| \exp[(\gamma_1 + \beta - \gamma)t] \quad \text{for} \quad t \geq 0,$$

and since the numbers $\gamma_1 > 0$ and $\beta > 0$ can be chosen arbitrarily small, estimate (29.51) coincides with estimate (29.16). The theorem is proved.

30. Fundamental Definitions and Theorems of Lyapunov's Second Method for Equations with Time Delay

Our general approach in this work is to consider as elements of the trajectory $x(x_0(\cdot), t_0, t)$ for systems of equations with time delay

$$(30.1) \qquad \frac{dx_i}{dt} = X_i(x_1(t + \vartheta), \cdots, x_n(t + \vartheta), t) \qquad (i = 1, \cdots, n),$$

a segment of these trajectories $x(x_0(\cdot), t_0, t + \vartheta)$ for $-h \leq \vartheta \leq 0$. The theorems of Lyapunov's second method can be carried over without change to equations with time delay. An extension of these theorems to equations with delay in the time t was carried out by Èl'sgol'c [43], who pointed out, however, that his formulations have limited value. The reason for this is that in the majority of cases the Lyapunov theorems do not have valid converses, and therefore the method does not necessarily have universal application as it does for ordinary differential equations.

Now let us consider one possible method of generalizing the use of Lyapunov functions for equations with time delay. We regard the elements of the trajectory as vector functions $\{x_i(x_0(\cdot), t_0, t + \vartheta)\}$, which form sections of the trajectory for $-h \leq \vartheta \leq 0$. Our approach is the natural one of replacing the Lyapunov function $v(x_1(\vartheta), \cdots, x_n(\vartheta), t)$, defined on the vector $\{x_1, \cdots, x_n\}$, by the functional $V(x_1(\vartheta), \cdots, x_n(\vartheta), t)$, defined over the vector function $\{x_1(\vartheta), \cdots, x_n(\vartheta)\}$. Once this is done, the fundamental theorems and definitions of Lyapunov's second method carry over very naturally to the functional V, and the theorems have valid converses. In accordance with the remark made in section 28 (pp. 131–32), we confine our discussions to theorems on stability. We can assume throughout this section that the functionals X_i in equation (30.1) satisfy the conditions formulated in section 27, pp. 127–28. We begin with a definition.

DEFINITION 30.1. *The functional* $V(x_1(\vartheta), \cdots, x_n(\vartheta), t)$, *defined for the vector functions* $\{x_1(\vartheta), \cdots, x_n(\vartheta)\}$, $-h \leq \vartheta \leq 0$, *for*

$$(30.2) \qquad\qquad ||x(\vartheta)|| < H, \quad t \geq 0,$$

is called positive-definite in the region (30.2) *if there is a continuous function* $w(r)$ *that satisfies the conditions*

$$(30.3) \qquad \begin{aligned} &V(x_1(\vartheta), \cdots, x_n(\vartheta), t) \geq w(||x(\vartheta)||^{(h)}), \\ &w(r) > 0 \quad \text{for} \quad r \neq 0. \end{aligned}$$

We say that the functional $V(x_1(\vartheta), \cdots, x_n(\vartheta), t)$ has an infinitely small upper bound in the space (30.2) if there is a continuous function $W(r)$ satisfying the conditions

(30.4) $V(x_1(\vartheta), \cdots, x_n(\vartheta), t) \leq W(|| x(\vartheta) ||^{(h)}),$ $W(0) = 0$.

Let us evaluate the functional $V(x_1(\vartheta), \cdots, x_n(\vartheta), t) = V(x(\vartheta), t)$ when $x(\vartheta)$ is the vector function $\{x_i(x_0(\vartheta_0), t_0, t + \vartheta)\}, -h \leq \vartheta \leq 0$, which defines an element (segment) of the trajectory corresponding to the moment of time $t \geq t_0$. The value of the functional is a function $v(t)$:

(30.5) $v(t) = V\left(x_1(x_0(\vartheta_0), t_0, t + \vartheta), \cdots, x_n(x_0(\vartheta_0), t_0, t + \vartheta), t\right).$

In the following pages we shall be interested in the size of the upper right derivative of the function $v(t)$, that is, the value of

$$\lim_{\Delta t \to +0} \sup \left(\frac{v(t + \Delta t) - v(t)}{\Delta t} \right)$$

$$= \lim_{\Delta t \to +0} \sup \left(\frac{V(x(x_0(\vartheta_0), t_0, t + \Delta t + \vartheta), t + \Delta t) - V(x(x_0(\vartheta_0), t_0, t + \vartheta), t)}{\Delta t} \right)$$

This value can be denoted by

(30.6) $\left(\lim_{\Delta t \to +0} \sup \frac{\Delta V}{\Delta t} \right)_{(30.1)}$,

where the index (30.1) indicates that the limit of $\Delta V / \Delta t$ is computed along a trajectory of the system (30.1). The value of this limit will be called negative-definite in the region (30.2) if we can find a continuous function $w_1(r)$ satisfying the conditions

(30.7)
$$\left(\lim_{\Delta t \to +0} \sup \frac{\Delta V}{\Delta t} \right)_{(30.1)} \leq -w_1(|| x(x_0(\vartheta_0), t_0, t + \vartheta) ||^{(h)}) ,$$

$$w_1(r) > 0 \qquad\qquad (t \geq 0, \; 0 < r < H) .$$

Now we can formulate a theorem corresponding to the Lyapunov theorem on asymptotic stability.

THEOREM 30.1. *If to a differential equation (30.1) with time delay there corresponds a functional $V(x(\vartheta), t)$ which is positive-definite in region (30.2)*[*] and has an infinitely small upper bound, and if the value (30.6) is negative-definite along equation (30.1) [in region (30.2)], then the null solution $x = 0$ of equation (30.1) is asymptotically stable. If, moreover, the condition*

(30.8) $\inf [V \; for \; || x(\vartheta) ||^{(h)} = H_1] > \sup [V \; for \; || x(\vartheta) ||^{(h)} = H_0]$

$$(H_0 < H_1 < H)$$

is satisfied, then the region of initial curves

(30.9) $|| x_0(\vartheta_0) ||^{(h)} < H_0$

lies in the region of attraction of the point $x = 0$.

---

[*] As with ordinary Lyapunov functions, we suppose that the functional V is continuous in $x(\vartheta)$ and t.

Remark. Our formulation of theorem 30.1 uses as the meaning of deriva-
tive dV/dt the value of $\lim \sup (\Delta V/\Delta t)$ [as $\Delta t \to +0$] along the integral curve.
Thus, since the bound we are here considering is only for $\Delta t \to +0$, it is
clear in particular that the trajectories of a system with time delay cannot
always be continued in the direction of decreasing t; that is, on the curves
$\{x_i(\vartheta)\}$ we cannot compute ΔV for $\Delta t < 0$. We restrict ourselves to the right-
hand upper bound for two reasons: first, to establish stability, it is sufficient
that the upper bound of the difference quotient $\Delta V/\Delta t$ exist on the right, as
the proof of the theorem establishes. Second, there is no simple formula
for the derivative dV/dt for equations with time delay comparable to formula
(1.5) for the derivative dv/dt in the case of ordinary equations. The need
to verify the existence of dV/dt would make it very difficult to investigate
actual equations, and would in fact restrict us to the small subclass of
equations that can be investigated by use of theorem 30.1.

PROOF OF THEOREM 30.1. Let a positive number $\varepsilon > 0$ be given. Then we
can choose $\delta > 0$ such that $w(\varepsilon) > W(\delta)$, in which case, by (30.3) and (30.4),
the inequality

$$(30.10) \qquad \sup [V \quad \text{for} \quad || x(\vartheta) ||^{(h)} = \delta] < \inf [V \quad \text{for} \quad || x(\vartheta) ||^{(h)} = \varepsilon]$$

is valid.

Let $|| x_0(\vartheta) ||^{(h)} < \delta$. Since the quantity (30.6) is negative-definite, the func-
tion $v(t)$ defined in (30.5) is a monotonically decreasing function of the time
t; that is, $v(t) < v(t_0)$. Therefore, for all $t \geq t_0$ the trajectory $x(x_0(\vartheta_0), t_0, t)$
lies in the region $|| x(x_0(\vartheta_0), t_0, t) || < \varepsilon$, since otherwise inequality (30.10)
would be violated. Thus the stability of the solution $x = 0$ in the sense of
definition 28.1(a) is proved.

Now let $x_0(\vartheta_0)$ be an initial vector function in the region (30.9). We consider again the fact that the function $v(t)$ (30.5) is monotonically decreasing.
Since $v(t) < v(t_0)$ for $t > t_0$, the trajectory $x(x_0(\vartheta_0), t_0, t)$ lies in the region
$|| x || < H_1$; otherwise, inequality (30.8) would not be satisfied.

Now let us take positive $\varepsilon > 0$, and let $\delta > 0$ be defined from ε as in the
first part of the proof. In the region $\delta \leq || x || \leq H_1$, the inequality

$$(30.11) \qquad \left(\lim_{\Delta t \to +0} \sup \frac{\Delta V}{\Delta t} \right)_{(30.1)} < -\alpha$$

is satisfied for some positive constant $\alpha > 0$. Thus the function $v(t)$ will
satisfy the inequality

$$(30.12) \qquad v(t) - v(t_0) \leq -\alpha(t - t_0)$$

for all those values of $t \geq t_0$ for which the segment $x(x_0(\vartheta_0), t_0, t)$ of the
trajectory lies in the region

$$(30.13) \qquad \delta \leq || x(x_0(\vartheta_0), t_0, t + \vartheta) ||^{(h)} \leq H_1 .$$

Indeed, if we assume that inequality (30.12) is violated for $t > t_1 \geq t_0$,
but valid for $t < t_1$, then the inequality

$$\left(\lim_{\Delta t \to +0} \sup \frac{\Delta v(t)}{\Delta t} \right)_{t=t_1} > -\alpha$$

will be satisfied. But this contradicts inequality (30.11). Now inequality (30.12) shows that the trajectory $x(x_0(\vartheta_0), t_0, t)$ cannot lie in the region (30.13) for $t \geq t_0$ on the segment $t_0 \leq t \leq t_0 + v(t_0)/\alpha$. Since $v(t_0)/\alpha \leq W(H_1)/\alpha$ by (30.3) and (30.4), we can see that there is some $t^* \in [t_0, t_0 + W(H_1)/\alpha]$ which satisfies the inequality

$$|| x(x_0(\vartheta_0), t_0, t^* + \vartheta) ||^{(h)} < \delta ;$$

and consequently, for all $t > t^*$ (and *a fortiori* for $t > t_0 + W(H_1)/\alpha$), we have

$$|| x(x_0(\vartheta), t_0, t + \vartheta) ||^{(h)} < \varepsilon .$$

But this inequality establishes the theorem, since the original number $\varepsilon > 0$ could be chosen arbitrarily.

Remark. Since neither $T = W(H_1)/\alpha$, nor α, nor the estimate of δ (as a function of ε) depends on the moment $t_0 < 0$ or on the value of the initial curve $x_0(\vartheta_0)$ in region (30.9), we can conclude that the solution $x = 0$ is not only asymptotically stable under the hypotheses of theorem 30.1, but also asymptotically stable uniformly in the coordinates $x_0(\vartheta_0)$ of region (30.9) and in the time $t_0 \geq 0$ in the sense of the following definition:

DEFINITION 30.2. *The solution $x = 0$ of the system of equations (30.1) with time delay is called uniformly asymptotically stable with respect to the time $t_0 \geq 0$ and with respect to the initial curve $x_0(\vartheta_0)$ in region (30.9) if it satisfies condition* (b) *of definition 28.1 and also satisfies the following conditions:* (i) *the number $\delta > 0$ of definition 28.1(a) may be chosen independent of $t_0 \geq 0$;* (ii) *for arbitrary $\eta > 0$, there exists a number $T(\eta)$ such that*

$$(30.14) \qquad || x(x_0(\vartheta_0), t_0, t + \vartheta) ||^{(h)} < \eta$$

holds for every $t \geq t_0 + T(\eta)$, independent of the choice of a piecewise-continuous initial curve $x_0(\vartheta_0)$ in region (30.9). We call this property B'.

This definition coincides with definition 5.1 (p. 26) for ordinary equations.

Now let us consider the possible converse of theorem 30.1. The proof of theorem 30.1 shows that the possibility of inverting this theorem is slim unless the stability is uniform with respect to $x_0(\vartheta_0)$ and t_0. We shall show that in this case theorem 30.1 does indeed have a valid converse. That is, for equations with time delay, theorem 30.1 is just as general as the theorem on asymptotic stability [114, p. 259] for ordinary differential equations.

THEOREM 30.2. *If the null solution $x = 0$ of the system (30.1) is asymptotically stable uniformly with respect to initial curves $x_0(\vartheta_0)$ and t_0 in region (30.9), that is, uniformly stable in the sense of definition 30.2 (property B'), then there is a functional $V(x(\vartheta), t)$, defined on the curves $|| x(\vartheta) ||^{(h)} < H_0$ for $t \geq 0$, positive-definite, having an infinitely small upper bound in this region, and having, moreover, a right derivative $dV/dt = \lim (\Delta V/\Delta t)$ for $\Delta t \to +0$ on the trajectory of system (30.1). The derivative dV/dt is a negative-definite* *functional in some region* $||x(\vartheta)||^{(h)} < H_0$, *and the functional* $V(x(\vartheta), t)$ *satisfies a Lipschitz condition with respect to* $x(\vartheta)$; *that is,*

$$(30.15) \qquad |V(x''(\vartheta), t) - V(x'(\vartheta), t)| \leq K_0 ||x''(\vartheta) - x'(\vartheta)||^{(h)} .$$

PROOF. To prove this theorem, we shall construct a Lyapunov function in the manner used by Massera [129] for ordinary differential equations. Let $\varphi(t)$ be a monotonically decreasing continuous function, satisfying the inequality

$$(30.16) \qquad ||x(x_0(\vartheta), t_0, t + \vartheta)||^{(h)} \leq \varphi(t - t_0) \quad \text{for} \quad t \geq t_0$$

on an arbitrary initial curve $||x_0(\vartheta_0)||^{(h)} < H_0$, and such that $\lim \varphi(t) = 0$ as $t \to \infty$. The existence of a function $\varphi(\tau)$ of this character follows from the fact that the solution is uniformly asymptotically stable in the sense of definition 30.2 (property B'). (Such a function $\varphi(\tau)$ can be found, for instance, by taking any continuous monotonically decreasing function that satisfies the inequality $\varphi(\tau) > \varepsilon$ for $\tau \in [T(\varepsilon), T(\varepsilon/2)]$.)

Malkin [124] gives the details of the construction of the function $\varphi(\tau)$ and also a proof of the existence of a function $G(\varphi)$ that is monotonically increasing and continuously differentiable, with the following properties:

$$(30.17) \qquad G(\varphi) > 0 \quad \text{for} \quad \varphi > 0 ,$$

$$(30.18) \qquad \int_0^\infty G(\varphi(\tau))\, d\tau = N < \infty ,$$

$$(30.19) \qquad \int_0^\infty G'(\varphi(\tau))e^{L\tau}\, d\tau = N_1 < \infty ,$$

$$(30.20) \qquad G'(\varphi(\tau))e^{L\tau} < N_2 \quad \text{for all} \quad \tau \geq 0 .$$

We shall now show that the functional

$$(30.21) \qquad V(x(\vartheta), t) = \int_t^\infty G(||x(x(\vartheta), t, \tau + \vartheta)||^{(h)})\, d\tau$$

$$+ \sup [G(||x(x(\vartheta), t, \tau + \vartheta)||^{(h)}) \quad \text{for} \quad t \leq \tau < \infty]$$

satisfies all the conditions of the theorem.

Since the integral (30.18) converges, it follows from inequality (30.16) that the integral in (30.21) converges for all $||x(\vartheta)||^{(h)} < H_0$. That is, the function V is defined in the region

$$(30.22) \qquad ||x(\vartheta)||^{(h)} < H_0 .$$

Since the function $G(\varphi)$ is positive for $\varphi > 0$, the first term in the right member of (30.21) is positive [for $||x(\vartheta)||^{(h)} \neq 0$], and the second is no less than $G(||x(\vartheta)||^{(h)})$; that is, the functional V satisfies the inequality

$$Vx((\vartheta), t) \geq G(||x(\vartheta)||^{(h)}) ,$$

and is therefore positive-definite.

Now let us show that the functional $V(x(\vartheta), t)$ has an infinitely small upper bound. First, we remark that according to the definition of asymptotic stability the relation

$$|| x(x_0(\vartheta_0), t_0, t + \vartheta) ||^{(h)} < H_1 \quad \text{for} \quad t \geq t_0 \quad \text{for all} \quad || x_0(\vartheta_0) ||^{(h)} \leq H_0$$

is valid in the region $|| x || < H_0$; thus

$$V(x(\vartheta), t) < \int_0^\infty G(\varphi(\tau)) \, d\tau + G(\varphi(0)) = N + G(\varphi(0)) = Q = \text{const.}$$

for all $|| x(\vartheta) ||^{(h)} \leq H_0$; that is, the functional $V(x(\vartheta), t)$ is uniformly bounded in the region (30.22). Consequently, to show that the functional V has an infinitely small upper bound, it would be enough to show that $V \to 0$ uniformly for $|| x(\vartheta) ||^{(h)} \leq H_0$. As a preliminary step in this, we show that the functional V satisfies the Lipschitz condition (30.15). We have

$$|V(x''(\vartheta), t) - V(x'(\vartheta), t)|$$
$$= \left| \int_t^\infty \left[G\left(|| x(x''(\vartheta), t, \tau + \vartheta) ||^{(h)} \right) - G\left(|| x(x'(\vartheta), t, \tau + \vartheta) ||^{(h)} \right) \right] d\tau \right.$$
$$\left. + \left[\sup\left(|| x(x''(\vartheta), t, \tau + \vartheta) ||^{(h)} \right) - \sup\left(|| x(x'(\vartheta), t, \tau + \vartheta) || \right) \right] \right|$$
$$\leq \int_t^\infty G'_\varphi\left(\sup\left\{ || x(x''(\vartheta), t, \tau + \vartheta) ||^{(h)}, || x(x'(\vartheta), t, \tau + \vartheta) ||^{(h)} \right\} \right)$$
$$\cdot \left(|| x(x''(\vartheta), t, \tau + \vartheta) - x(x'(\vartheta), t, \tau + \vartheta) ||^{(h)} \right) d\tau$$
$$+ \sup \left| G\left(|| x(x''(\vartheta), t, \tau + \vartheta) ||^{(h)} \right) - G\left(|| x(x'(\vartheta), t, \tau + \vartheta) ||^{(h)} \right) \right|.$$

At this point, we use inequality (27.12) and the second theorem of the mean to obtain the further inequality

$$|V(x''(\vartheta), t) - V(x'(\vartheta), t)|$$
$$\leq \left[\int_t^\infty G'(\varphi(\tau - t)) e^{L(\tau - t)} \, d\tau + \sup\left(G'_\varphi(\varphi(\tau - t)) e^{L(\tau - t)} \quad \text{for} \quad t < \tau \leq \infty \right) \right]$$
$$\cdot || x''(\vartheta) - x'(\vartheta) ||^{(h)}$$
$$= (N_1 + N_2) || x''(\vartheta) - x'(\vartheta) ||^{(h)},$$

which shows that relation (30.15) holds and that there is indeed an infinitely small upper bound. Setting $x''(\vartheta) = x(\vartheta)$ and $x'(\vartheta) = 0$, we then have

$$V(x(\vartheta), t) \leq K || x(\vartheta) ||^{(h)}.$$

Finally, we shall show that the functional $V(x(\vartheta), t)$ is continuous in $x(\vartheta)$ and t. From the proof of inequality (30.15), it is sufficient to show merely that $V(x(\vartheta), t)$ is continuous in t for a fixed vector function $x(\vartheta)$.

The first step in the demonstration will be to show that the function $V(x(\vartheta), t)$ has right upper and lower derivatives that are uniformly bounded as long as the system of equations (30.1) is satisfied. We have (for $t = t_0$)

$$(30.23) \quad \left(\lim_{\Delta t \to +0} \sup \frac{\Delta V}{\Delta t} \right)_{t=t_0} = \frac{d}{dt} \left[\int_t^\infty G\Big(|| x(x(x_0(\vartheta), t_0, t), t, \tau + \vartheta) ||^{(h)} \Big) d\tau \right]_{t=t_0}$$

$$+ \lim_{\Delta t \to +0} \sup \left\{ \sup \left[G\Big(|| x(x_0(\vartheta), t_0 + \Delta t, \tau + \vartheta) ||^{(h)} \Big) \quad \text{for} \quad t_0 + \Delta t \leq \tau < \infty \right] \right.$$

$$\left. - \sup \left[G\Big(|| x(x_0(\vartheta), t_0, \tau + \vartheta) ||^{(h)} \Big) \quad \text{for} \quad t_0 \leq \tau < \infty \right] \right\} .$$

The second term in the right member of this equation is nonnegative. Thus

$$(30.24) \quad \left(\lim_{\Delta t \to +0} \sup \frac{\Delta V}{\Delta t} \right)_{t=t_0} \leq -G\Big(|| x(x_0(\vartheta), t_0, t_0 + \vartheta) ||^{(h)} \Big) = -G(||x(\vartheta)||^{(h)}) .$$

This relation shows that the derivative of the first term in the right member of equation (30.23) simply reduces to the right derivative of the integral with respect to the variable lower limit t.

From relation (30.24) it follows that the value

$$\left(\lim_{\Delta t \to +0} \sup \frac{\Delta V}{\Delta t} \right)_{t=t_0}$$

is indeed a negative-definite functional. Now let us compute the lower right derivative $\lim \inf \Delta V / \Delta t$ for $t \to +0$. The computation is similar to that of (30.23) and we obtain analogously (for $t = t_0$)

$$\left(\lim_{\Delta t \to +0} \inf \frac{\Delta V}{\Delta t} \right)_{t=t_0} = -G(|| x_0(\vartheta) ||^{(h)})$$

$$+ \lim \inf \left\{ \sup \left[G\Big(|| x(x(x_0(\vartheta), t_0, t), t, \tau + \vartheta) ||^{(h)} \Big) \quad \text{for} \quad t \leq \tau < \infty \right] \right.$$

$$\left. - \sup \left[G\Big(|| x(x_0(\vartheta), t_0, \tau + \vartheta) ||^{(h)} \Big) \quad \text{for} \quad t_0 \leq \tau < \infty \right] \right\}$$

$$\geq -G(|| x_0(\vartheta) ||^{(h)}) - G'_\varphi \Big(|| x(x_0(\vartheta), t_0, \tau + \vartheta) ||^{(h)} \Big)$$

$$\cdot \sup \Big(| X_i(x_0(\vartheta), t_0) |, \quad i = 1, \cdots, n \Big) \geq -Q = \text{const.}$$

Thus we see that along the trajectory $x(x(\vartheta), t_0, t)$ the functional $V(x(\vartheta), t)$ has uniformly bounded upper and lower right derivatives in the region $|| x(\vartheta) ||^{(h)} < H_0$. For fixed $t > t_0$ and a fixed curve $x_0(\vartheta)$, we compute as follows:

$$| V(x_0(\vartheta), t_0) - V(x_0(\vartheta), t) |$$

$$\leq | V(x_0(\vartheta), t_0) - V(x(x_0(\vartheta), t_0, t), t) | + | V(x_0(\vartheta), t) - V(x(x_0(\vartheta), t_0, t), t) |$$

$$\leq Q_1 \Delta t + \sup [| X_i(x(\vartheta), t) | \quad \text{for} \quad || x(\vartheta) ||^{(h)} < H; i = 1, \cdots, n] K \Delta t;$$

that is,

$$(30.25) \qquad | V(x_0(\vartheta), t_0) - V(x_0(\vartheta), t) | \leq Q_2 \Delta t \qquad (Q_2 = \text{const.}) .$$

From inequalities (30.15) and (30.25), we can see that the functional $V(x(\vartheta), t)$ depends continuously on t in the following sense: For an arbitrary $\varepsilon > 0$, we can find a $\delta > 0$ such that

$$| V(x_0(\vartheta), t_0) - V(x(\vartheta), t) | < \varepsilon$$

whenever

$$|t - t_0| < \delta \quad \text{and} \quad || x(\vartheta) - x_0(\vartheta) ||^{(h)} < \delta \, .$$

Thus we have shown that the functional V satisfies all the conditions of the theorem. The theorem is proved.

It has been shown that a necessary and sufficient condition for the existence of the functional V of theorem 30.2 is that the null solution be asymptotically stable uniformly for a given initial curve $x_0(\vartheta)$ and initial moments of time $t_0 \geqq 0$. In this respect, the situation is similar to that in the case of ordinary differential equations. If the right member of equation (30.1) is a periodic function of time with period θ (or is independent of the time t), then the null solution $x = 0$ is always uniformly asymptotically stable in the sense of property B′ (definition 30.2, p. 146).

We shall now prove this assertion. Let the null solution $x = 0$ be an asymptotically stable solution of equation (30.1) and let the right member of this equation be a periodic function of the time with period θ. Furthermore, let the initial curve

(30.26) $$|| x_0(\vartheta) ||^{(h)} \leq H_0$$

lie in a region containing the point $x = 0$. We shall show first that if $\varepsilon > 0$ is an arbitrary positive number, we can find a number $\delta > 0$ such that $|| x(x_0(\vartheta_0), t_0, t) || < \varepsilon$ for all $t \geqq t_0$, provided $|| x_0(\vartheta_0) ||^{(h)} \leq \delta$. Since X_i is periodic, it is certainly sufficient to demonstrate this assertion for values of t_0 in the interval $[0, \theta)$. The definition of stability shows that for a given positive $\varepsilon > 0$, we can find a number $\delta(\varepsilon, \theta) > 0$ such that

$$|| x(x_0(\vartheta_0), \theta, t + \vartheta) ||^{(h)} < \varepsilon \quad \text{for all} \quad t \geqq \theta \, ,$$

provided that

$$|| x_0(\vartheta_0) ||^{(h)} \leq \delta(\varepsilon, \theta) \, .$$

On the other hand, inequality (27.12) shows that

$$|| x(x_0(\vartheta_0), t_0, \theta + \vartheta) ||^{(h)} \leq || x_0(\vartheta_0) ||^{(h)} e^{L\theta}$$

for all $t_0 \in [0, \theta)$.

Thus the truth of our assertion can be seen if we simply define the number δ by $\delta = \delta(\varepsilon, \theta) e^{-L\theta}$.

Now we are able to demonstrate that the null solution is uniformly asymptotically stable in the sense of definition 30.2 (property B′). Let us suppose the contrary. Then there is a sequence of curves $x_0^{(\nu)}(\vartheta)$ and initial moments of time $t_0^{(\nu)} \in [0, \theta)$, $\nu = 1, 2, \cdots$, so arranged that there are points t_ν on the trajectory $x(x_0^{(\nu)}(\vartheta_0), t_0^{(\nu)}, t)$ that satisfy the inequality

$$\| x(x_0^{(\nu)}(\vartheta_0), t_0^{(\nu)'}, t_\nu + \vartheta) \|^{(h)} > \varepsilon \,,$$

where, furthermore, $| t^{(\nu)} - t_\nu | \to \infty$ for $\nu \to \infty$ and

(30.27) $$\| x_0^{(\nu)}(\vartheta_0) \|^{(h)} < H_0 \, e^{-2L\theta} \,.$$

Since the functionals X_i are periodic, we can assume that the numbers $t_0^{(\nu)}$ lie in the interval $[0, \theta)$. The sequence $x(x_0^{(\nu)}, t_0^{(\nu)}, 2\theta + \vartheta)$ of functions of the argument ϑ is uniformly bounded, continuous, and equicontinuous. Therefore, we can choose a subsequence that is uniformly convergent.

To avoid the necessity of using a double subscript, let us assume that the sequence we have is already this subsequence. Let $x_0(\vartheta)$ be the limit function of this sequence $x(x_0^{(\nu)}(\vartheta_0), t_0^{(\nu)}, 2\theta + \vartheta)$. Let us consider the trajectory $x(x_0(\vartheta_0), 2\theta, t)$. Since the null solution $x = 0$ is asymptotically stable with respect to the initial curves (30.26), our choice of the initial curves (30.27) makes it clear that we can assume that the curve $x_0(\vartheta)$ lies in a region containing the point $x = 0$. Therefore, corresponding to the number $\delta > 0$ we can find a number T such that the inequality

$$\| x(x_0(\vartheta_0), 2\theta, T + \vartheta) \|^{(h)} < \frac{\delta}{2}$$

holds, and furthermore, from inequality (27.12) we can find a number N such that the inequality

$$\| x(x_0^{(\nu)}(\vartheta_0), t_0^{(\nu)}, -t_0^{(\nu)} + 2\theta + T + \vartheta) \|^{(h)} < \delta$$

holds for all $\nu > N$. But we have shown above that under these circumstances the inequality

$$\| x(x_0^{(\nu)}(\vartheta_0), t_0^{(\nu)}, -t_0^{(\nu)} + 2\theta + t + \vartheta) \|^{(h)} < \varepsilon$$

holds for all $t > T$. This contradicts the choice of the subsequences $x_0^{(\nu)}(\vartheta_0)$ and $t_0^{(\nu)}$. This contradiction shows that the solution $x = 0$ is uniformly asymptotically stable with respect to the initial curves (30.27).

Thus, if X_i is a periodic functional, a necessary and sufficient condition for the null solution $x = 0$ to be asymptotically stable is that there should exist a functional $V(x(\vartheta), t)$, defined in the neighborhood of the solution $x = 0$, that satisfies the conditions of theorem 30.1.

31. Some Sufficient Conditions for Asymptotic Stability of Equations with Time Delay

The proof of theorem 30.1 in the preceding section gives conditions that are sufficient and (under definite restrictions) necessary for asymptotic stability of equations with time delay. However, in solving an actual problem, we would only rarely be able to construct a functional that satisfies the conditions of this theorem. In this section we shall give a certain generalization of theorem 30.1 that makes it possible, under definite conditions, to lift some of the requirements on the functional V.

THEOREM 31.1. *If a functional $V(x(\vartheta), t)$ exists, defined on curves $\{x_i(\vartheta)\}$,*

$-h \leqq \vartheta \leqq 0$, $i = 1, \cdots, n$, *having an infinitely small upper bound, and satis-fying the conditions*

(i)

(31.1)
$$V(x(\vartheta), t) \geqq f(|| \, x(\tau) \, ||_{(-h_1 \leqq \tau \leqq 0)}) \, ,$$

where $f(r)$ is a function that is positive and continuous for $r \neq 0$, and

$|| \, x(\tau) \, ||_{(-h_1 \leqq \tau \leqq 0)}$ *means* $\sup [|x_i(\tau)| \; for \; -h_1 \leqq \tau \leqq 0, \; i = 1, \cdots, n]$,

$$h_1 \leqq h, \; x_i(\tau) = x_i(\vartheta) \; for \; \tau = \vartheta \, ;$$

(ii)

(31.2)
$$\limsup_{\Delta t \to +0} \left(\frac{\Delta V}{\Delta t} \right) \leqq -\varphi(|| \, x(\tau) \, ||_{(-h_3 \leqq \tau \leqq -h_2)}) \, ,$$

whenever $x(t)$ satisfies equation (30.1), where $\varphi(r)$ is also positive and continuous for $r \neq 0$,

$|| \, x(\tau) \, ||_{-h_3 \leqq \tau \leqq -h_2}$ *means* $\sup [| \, x_i(\tau)| \; for \; -h_3 \leqq \tau \leqq -h_2, i = 1, \cdots, n]$,

$$h \leqq h_3 \leqq h_2 \leqq 0 \, ;$$

then the solution $x = 0$ is asymptotically stable.

PROOF. Let a positive number $\varepsilon > 0$ be given. If there exists a number $\delta > 0$ such that the inequality

$$\sup [V(x_0(\vartheta_0), t_0) \; for \; || \, x_0(\vartheta_0) \, ||^{(h)} \leqq \delta] < \inf [V(x(\vartheta), t) \; for \; || \, x(\vartheta) \, ||_{(-h_1 \leqq \vartheta \leqq 0)}]$$

holds, then since V is a nonincreasing function of the time t, the trajectory $x(x_0(\vartheta_0), t_0, t)$ lies in the region $|| \, x(\vartheta) \, ||_{(-h_1 \leqq \vartheta \leqq 0)} < \varepsilon$. This proves that the solution $x = 0$ is stable.

Let us show that the null solution $x = 0$ is asymptotically stable. If we assumed the contrary, there would have to be a trajectory $x(x_0(\vartheta_0), t_0, t)$ such that its initial curve would satisfy the inequality

$$|| \, x_0(\vartheta_0) \, ||^{(h)} < \delta \, ,$$

and at the same time, there would have to be a positive number $\eta > 0$ such that $|| \, x(x_0(\vartheta_0), t_0, t) \, ||^{(h)} > \eta$ for all $t \geqq t_0$. In other words, we are assuming that there exists a sequence of numbers t_ν, $\nu = 1, 2, \cdots$, that increases without bound and is such that the inequality

(31.3)
$$|| \, x(x_0(\vartheta_0), t_0, t_\nu) \, || > \eta > 0$$

holds. The trajectory $x(x_0(\vartheta_0), t_0, t)$ lies in the region $|| \, x(\vartheta) \, || < \varepsilon$ for $t > t_0$, as proved above, where the speed of the point $\{x_i(t)\}$ in the phase space $\{x_i\}$ is uniformly bounded by some constant P. Therefore, there is a fixed number $\beta > 0$ such that

$$|| \, x(x_0(\vartheta_0), t_0, t) \, || > \frac{\eta}{2} \; for \; t_\nu - \beta \leqq t \leqq t_\nu + \beta \, .$$

Thus, the inequality

$$\lim_{\Delta t \to +0} \sup \left(\frac{\Delta V}{\Delta t}\right) < -\varphi\left(\frac{\eta}{2}\right)$$

is valid for the interval of time Δt_ν, where

$$\Delta t_\nu = (-h_3 - \beta + t_\nu \leq t \leq t_\nu + \beta + h_2) \,,$$

by condition (31.2).

Since the segments Δt_ν can be taken as nonoverlapping, the following inequality is valid along a trajectory

$$V(x(x_0(\vartheta_0), t_0, t), t) - V(x_0(\vartheta_0), t_0) \leq -\varphi\left(\frac{\eta}{2}\right) \sum_{\nu=1}^{N} \Delta t_\nu \,.$$

Here the sum is taken for all segments Δt_ν that lie wholly within the interval $[t_0, t]$. Since the sum of all such segments Δt_ν is supposed to have increased without bound as $t \to \infty$ ($N \to \infty$), then according to our assumptions, it follows that $\lim V(x(x_0(\vartheta_0), t_0, t), t) = -\infty$ for $t \to \infty$, which is impossible. This contradiction proves the theorem.

Remark. In case either $h_1 = 0$ or $h_2 = h_3$ the estimate for the functional becomes an estimate for an ordinary function. Moreover, if $h_1 = 0$, the functional V may be a simple function $v(x, t)$.

Now let us consider the case when the right members X_i are periodic functions of time and all have the same period θ (or are in fact independent of time). In this case condition (31.2) can be somewhat weakened.

THEOREM 31.2. *Suppose that there is a functional V that*
(i) *is periodic in time t (or independent of the time t),*
(ii) *has an infinitely small upper bound, and*
(iii) *satisfies condition (31.1).*
Suppose further that
(iv) $\lim \sup_{\Delta t \to +0}(\Delta V / \Delta t)$ *has a nonpositive value along a trajectory;*
(v) *in some neighborhood of $x = 0$ there is no trajectory (except the null solution itself) along which the relation $\lim \sup (\Delta V / \Delta t) = 0$ holds for all $t \geq t_0$.*
Then the solution $x = 0$ is asymptotically stable.

PROOF. To prove theorem 31.2, it is sufficient to establish that the point $x = 0$ is asymptotically stable; we already know it is stable. If we suppose the opposite, then it follows just as it did in the proof of theorem 31.1 that there is a trajectory $x(x_0(\vartheta_0), t_0, t)$ that does not lie close to the point $x = 0$ for $|| x_0(\vartheta_0) ||^{(h)} < \delta$ as the time increases; that is, we assume that

$$|| x(x_0(\vartheta_0), t_0, t_0 + \nu\theta + \vartheta) ||^{(h)} > \eta \qquad (\nu = 1, 2, \cdots) \,.$$

The uniformly bounded sequence of equicontinuous functions

$$x(x_0(\vartheta_0), t_0, t_0 + \nu\theta + \vartheta) = y_\nu(\vartheta)$$

has at least one limit element $y_0(\vartheta)$. But the nonincreasing function

$$V(x(x_0(\vartheta_0), t_0, t), t) = v(t)$$

has a limit as $t \to \infty$. Calling this limit $v_\infty \neq 0$, it would follow from the continuity of the functional V that the equation

$$V(y_0(\vartheta), t_0) = v_\infty$$

must hold.

Let us consider the trajectory $x(y_0(\vartheta_0), t_0, t)$. The hypotheses of the theorem show that the function $V(x(y_0(\vartheta_0), t_0, t), t)$ is not identically constant for $t \geq t_0$. Thus there must be a value $t = t_0 + k\theta$ such that

$$V\Big(x(y_0(\vartheta_0), t_0, t_0 + k\theta), t_0\Big) < V(y_0(\vartheta), t_0) = v_\infty \,.$$

But $y_0(\vartheta)$ is a limit element of the sequence $x(x_0(\vartheta_0), t_0, t_0 + \nu\theta)$, and therefore, if ν is taken sufficiently great, the inequality

$$V\Big(x(x_0(\vartheta_0), t_0, t_0 + \nu\theta + k\theta), t_0 + \nu\theta + k\theta\Big) < v_\infty$$

must hold. Since this is impossible (by the definition of v_∞) the theorem is established.

Remark. Theorem 31.2, which is here proved for equations with time delay, corresponds to theorem 14.1, which we proved for ordinary differential equations.

To construct concrete functionals, it is sometimes convenient to make use of the norm defined by formula (27.6):

$$(31.4) \qquad || x(\vartheta) ||_2^{(h)} = \left(\int_{-h}^{0}\left[\sum_{i=1}^{n} x_i^2(\vartheta)\right]d\vartheta\right)^{1/2} \,.$$

The definition of stability given above used the metric defined by the norm (27.5). The metric defined by the norm (31.4) can be used in a corresponding manner.

Here we shall give this foundation for equations with delay in the time t. We consider the equation

$$(31.5) \qquad \frac{dx_i}{dt} = X_i(x_1(t - h_{i1}), \cdots, x_n(t - h_{in}); x_1(t), \cdots, x_n(t), t)$$

$$(i = 1, \cdots, n; \quad 0 \leq h_{ij}(t) \leq h) \,,$$

where the right members $X_i(y_1, \cdots, y_n; x_1, \cdots, x_n, t)$ are continuous functions of their arguments and are defined for

$$(31.6) \qquad |x_i| < H, \qquad |y_i| < H \qquad (i = 1, \cdots, n) \,.$$

We suppose further that these functions satisfy a Lipschitz condition with respect to the x_j and y_j in the region (31.6),

$$| X_i(y_1'', \cdots, y_n''; x_1'', \cdots, x_n'', t) - X_i(y_1', \cdots, y_n'; x_1', \cdots, x_n', t) |$$

$$\leq L\Big(\sum_{j=1}^{n} | x_j'' - x_j' | + \sum_{j=1}^{n} | y_j'' - y_j' |\Big) \,.$$

THEOREM 31.3. *Suppose there is a functional $V(x(\vartheta), t)$ that satisfies the conditions*

(31.7) $$| V(x(\vartheta), t) | \leq W_1(\| x(0) \|) + W_2(\| x(\vartheta) \|_2^{(h)}) ,$$

(31.8) $$V(x(\vartheta), t) \geq w(\| x(0) \|) ,$$

(31.9) $$\lim_{\Delta t \to +0} \sup \left(\frac{\Delta V}{\Delta t} \right) \leq -\varphi(\| x(0) \|) ,$$

where $W_1(r)$ and $W_2(r)$ are functions that are continuous and monotonic for $r \geq 0$, and $W_1(0) = W_2(0) = 0$; $w(r)$ is a function that is continuous and positive for $r \neq 0$; and $\varphi(r)$ is a function that is continuous and positive for $r \neq 0$. Then the null solution $x = 0$ of equations (31.5) is asymptotically stable.

Remark. If X_i and $h_{ij}(t)$ are periodic functions of the time t, all with period θ (or if X_i and h_{ij} are independent of the time t), then condition (31.9) may be replaced by the following weaker hypothesis: A sufficient condition that

$$\lim \sup (\Delta V / \Delta t) \quad \text{for} \quad \Delta t \to +0$$

should be nonpositive along a trajectory is that the equation

$$\lim \sup (\Delta V / \Delta t) = 0 \quad \text{for} \quad \Delta t \to +0$$

be valid for all $t \geq t_0$ only along the trajectory $x = 0$.

PROOF OF THE THEOREM. Here it is sufficient to prove only stability of the solution $x = 0$, since asymptotic stability is a necessary consequence of stability, as can be proved by mimicking the proofs of theorems 31.1 and 31.2. Let us suppose then that the positive number $\varepsilon > 0$ is given. We can find a number $\delta > 0$ such that the inequality

$$W_1(\delta) + W_2(\delta n h) < w(\varepsilon)$$

is satisfied. Then, according to inequalities (31.7) and (31.8), the inequality

(31.10) $$V(x_0(\vartheta), t_0) < w(\varepsilon) \quad \text{for} \quad \| x_0(\vartheta) \|^{(h)} < \delta$$

is valid. But the function

$$V(x x_0((\vartheta_0), t_0, t), t) = v(t)$$

does not increase along a trajectory (31.5) according to the hypothesis of the theorem. Therefore, we can conclude from inequality (31.10) that

$$V(x(x_0(\vartheta_0), t_0, t + \vartheta), t) < w(\varepsilon) \quad \text{for all} \quad t \geq t_0 .$$

Thus, by inequality (31.8) we have the relation

(31.11) $$\| x(x_0(\vartheta_0), t_0, t) \| < \varepsilon .$$

Inequality (31.11) shows that the solution $x = 0$ is stable, and the theorem is proved.

Now let us give a sufficient condition for asymptotic stability of the null solution $x = 0$ for equations with time delay.

First we make an auxiliary notational agreement. The symbol $N(H_0, t_0, t)$, for $t \geq t_0$, is any family of curves $\{x_i(\vartheta)\}$, $i = 1, \cdots, n$, $-h \leq \vartheta \leq 0$, that contains all the solutions $x(x_0(\vartheta_0), t_0, t + \vartheta)$ (considered as functions of the argument ϑ) which correspond to all possible initial curves $x_0(\vartheta_0)$ that are piecewise-continuous for $t_0 \geq 0$ and that lie in the region

$$(31.12) \qquad || x_0(\vartheta_0) ||^{(h)} \leq H_0 \qquad (H_0 = \text{const.}) .$$

LEMMA 31.1. *Let H be such that all solutions of equation (30.1) which correspond to all initial curves $x_0(\vartheta_0)$ in the region (31.12) lie, for $t_0 \leq t < \infty$, in the region*

$$(31.13) \qquad || x(\vartheta) || < H .$$

Suppose there is a function $v(x_1, \cdots, x_n, t)$ that is uniformly bounded for $|x_i| < H$, $t \geq 0$. Suppose further that the function v has the property that for arbitrary $\gamma > \gamma_0$ there are numbers $\alpha(\gamma) > 0$, $T(\gamma) > 0$ such that the inequality

$$(31.14) \qquad \left(\lim \sup_{\Delta t \to +0} \frac{\Delta v}{\Delta t} \right)_{(30.1)} < -\alpha(\gamma) \quad for \quad t \geq t_0 + T(\gamma)$$

is satisfied for all curves $x(\vartheta) \in N(H_0, t_0, t)$ that satisfy the inequalities

$$(31.15) \qquad v(x_1(0), \cdots, x_n(0), t) \geq \gamma .$$

Then the relation

$$(31.16) \qquad \lim \sup v(x(x_0(\vartheta_0), t_0, t), t) \leq \gamma_0 \quad for \quad t \to \infty$$

holds along the trajectories of (30.1) uniformly, independent of the initial curve in the region (31.12).

PROOF. Let the number $t_0 \geq 0$ be fixed. Suppose that along some trajectory $x(x_0(\vartheta_0), t_0, t)$, the inequality $v(x(x_0(\vartheta_0), t_0, t^*), t^*) \leq \gamma$ holds for $t^* \geq t_0 + T(\gamma)$. Then this inequality must also hold for all $t \geq t^*$. Indeed, if we suppose on the contrary that the inequality is invalid for $t^{**} > t^*$, then we can assert that at the moment $t = t^{**}$ there is a number $\tau > 0$ such that on the one hand $v(x(x_0(\vartheta_0), t_0, t), t) > \gamma$ for $t \in [t^{**}, t^{**} + \tau]$ (if $\tau > 0$ is a sufficiently small number), and on the other hand the inequality

$$v(x(x_0(\vartheta_0), t_0, t), t) < \gamma \quad for \quad t \in (t^{**}, t^{**} + \tau)$$

holds [in view of relations (31.14) and (31.15)]. This contradiction establishes our assertion.

Now let γ be fixed, $\gamma > \gamma_0$. Let $x(x_0(\vartheta_0), t_0, t)$ be some trajectory that satisfies inequality (31.15) for $t > t_0 + T(\gamma)$. According to the hypothesis of the theorem, the function $v(x(x_0(\vartheta_0), t_0, t), t)$ is a monotonically decreasing function of the time for all $t > t_0 + T(\gamma)$, and since $v(x(x_0(\vartheta_0), t_0, t), t)$ satisfies inequality (31.15), we obtain the relation

$$v\Big(x(x_0(\vartheta_0), t_0, t), t \Big) - v\Big(x(x_0(\vartheta_0), t_0, t_0 + T(\gamma)), t_0 + T(\gamma) \Big)$$

$$\leq -\alpha(\gamma)(t - t_0 - T(\gamma)) .$$

Therefore, we see that all these trajectories come into the region $v \leq \gamma$ for $t > t_0 + T(\gamma) + 2\bar{V}/a(\gamma)$, where $\bar{V} = \sup v$. Since the number $\gamma > \gamma_0$ can be taken arbitrarily close to the number γ_0, the lemma is proved.

Now we can state a theorem that is of fundamental importance for the application of Lyapunov functions to equations with time delay [56]. We call attention to the fact that a similar theorem was proved by Razumihin [156].

THEOREM 31.4.   *A sufficient condition for the null solution $x = 0$ of equation (30.1) to be asymptotically stable and the initial curves (31.12) to lie in the region of attraction of the point $x = 0$ is the existence of a function $v(x_1, \cdots, x_n, t)$ with the following properties:*

(i)   *v is positive-definite;*

(ii)   *v admits an infinitely small upper bound;*

(iii)   *there exist numbers H_0, H_1, H ($H_0 < H_1 < H$) such that*

(31.17) $\inf [v(x, t) \text{ for } \|x\| = H_1] > \sup [v(x, t) \text{ for } \|x\| \leq H_0]$;

(iv)   *along an arbitrary trajectory of the system (30.1) v satisfies the inequality*

$$\left(\frac{dv}{dt}\right)_{dt=+0} < -\varphi(\|x(t)\|) ,$$

and, for initial curves $x(\vartheta)$, the inequality

(31.18) $v(x_1(\xi), \cdots, x_n(\xi), \xi) < f(v(x_1(t), \cdots, x_n(t), t))$

holds for $t - h \leq \xi < t$, where $f(r)$ is some continuous function satisfying the inequality

$f(r) > r \text{ for } r \neq 0, \qquad f(r'') - f(r') > 0 \text{ for } r'' > r' .$

PROOF.   Let us show that for an arbitrary initial curve $x_0(\vartheta_0)$ which lies in the region (31.12), the inequality

(31.19) $\|x(x_0(\vartheta_0), t_0, t)\| \leq H_1$

holds for all $t \geq t_0$. For if there is a $t \geq t_0$ for which this assertion is not valid, then there is a t^* such that $t^* > t_0$ and for which the inequality

$$v(x(x_0(\vartheta_0), t_0, t), t) \leq v_1$$

is first violated, where $v_1 = \inf [v(x, t) \text{ for } \|x\| = H_1]$. Now we have $v(x(x_0(\vartheta_0), t_0, t), t) < v_1$ for $t < t^*$, since the relation $dv/dt < 0$ holds at $t = t^*$, according to the hypotheses of the theorem. This contradicts the supposition that (31.19) is violated. That is, if $\|x_0(\vartheta_0)\|^{(h)} \leq H_0$, the hypotheses of the theorem have as a consequence the fact that the trajectories $x(x_0(\vartheta_0), t_0, t)$ lie in $\|x\| < H$ for all $t \geq t_0$.

Lemma 31.1 shows that to complete the proof of the theorem it is sufficient to establish the fact that for every $\gamma > 0$ there is a number $T(\gamma) > 0$ such that the family of curves satisfying condition (31.18) for $t > t_0 + T(\gamma)$ includes all integral curves $\{x_i(t)\} = \{x_i(x_0(\vartheta_0), t_0, t)\}$ for which the inequality

$v(x(x_0(\vartheta_0), t_0, t), t) \geqq \gamma$ holds. The assertion is indeed true when $\gamma^* = v_1 + 1$. For we showed above that the integral curves $x(x_0(\vartheta_0), t_0, t)$ do not satisfy the inequality $v(x(x_0(\vartheta_0), t_0, t), t) \geqq v_1 + 1$ for $|| x_0(\vartheta_0) ||^{(h)} \leqq H_0$. Let $\gamma_0 > 0$ be the lower bound of all values of γ for which our assertion is valid. Now, $f(r)$ has the property that we can find a number $\eta > 0$, $\eta < \gamma_0$, such that the inequality $f(r) - r > 2\eta$ holds for $\gamma_0 - \eta \leqq r \leqq \gamma_0 + \eta$. Furthermore, lemma 31.1 shows that there is a number τ such that for all $t > t_0 + \tau$ the inequality $v(x(x_0(\vartheta_0), t_0, t), t) < \gamma_0 + \eta$ is satisfied for all initial curves $|| x_0(\vartheta_0) ||^{(h)} \leqq H_0$.

Therefore, the integral curves that satisfy the inequality

$$v(x(x_0(\vartheta_0), t_0, t), t) \geqq \gamma_0 - \eta \quad \text{for} \quad t > t_0 + \tau + h$$

satisfy the inequality

$$v(x(x_0(\vartheta_0), t_0, \xi), \xi) < \gamma_0 + \eta < f\Big(v(x(x_0(\vartheta_0), t_0, t), t) \Big) \qquad (\xi \in [t - h, t]) \,.$$

Thus, the number $\tau + h$ is such that, on the curves for which the relation

$$v(x(x_0(\vartheta_0), t_0, t), t) \geqq \gamma_0 - \eta \quad \text{for} \quad t > t_0 + \tau + h$$

holds, condition (31.14) of the lemma is satisfied. This contradicts the assumption that the number $\gamma_0 > 0$ is a lower bound of all the numbers γ for which the hypotheses of the lemma are satisfied, and this contradiction establishes the theorem.

NOTES. R. Driver [39b] has extended the above methods to cases in which the delay interval h [equation (27.1)] is a function of t. It is easily checked that if $0 < \lambda < 1$, the null solution of $x'(t) = x(\lambda t)$ is unstable on $[0, \infty)$. This book does not consider this equation; Driver does. Some Russian authors and Hale [71a] have considered similar equations—this fact is mentioned on p. 126 of this work. The interest of the work of Driver and Hale lies in their systematic extension of (27.1) to the case in which the functional in the right member of (27.1) depends on values of $x(t)$ over variable intervals; i.e., the right member of equation (27.2) is a functional, and not merely a function.

Recently Levin and Nohel [111c], [111d], [111e] have studied global behavior of solutions of a class of nonlinear Volterra equations and delay equations by using suitable Lyapunov functionals. As is to be expected, these do not satisfy the stability criteria of sections 30 and 31 above.

Consider first the equation

$$(31.20) \qquad \frac{dx}{dt} = -\int_0^t a(t - \tau) g(x(\tau)) \, d\tau \qquad (t \geqq 0) \,,$$

where

   (i) $a(t)$ is completely monotonic on $0 \leqq t < \infty$,

   (ii) $g(x) \in C(-\infty, \infty)$, $\quad xg(x) > 0 \; (x \neq 0)$, $\quad G(x) = \int_0^x g(\xi) \, d\xi \to \infty \; (|x| \to \infty)$.

(A special case occurs in certain reactor dynamics problems; the linear problem is considered in [111].) By using the Lyapunov functional

$$V(t) = G(u(t)) + \frac{1}{2} \int_0^t \int_0^t a(\tau + s) g(u(t - \tau)) g(u(t - s)) \, d\tau \, ds \, ,$$

where $u(t)$ is any solution of (31.20) for $t \geq 0$, together with S. Bernstein's representation [73, pp. 588, 592] of $a(t)$ as a Laplace-Stieltjes transform $\int_0^\infty e^{-st} d\alpha(s)$, the following result is obtained in [111c]:

THEOREM 31.5. *Let* (i), (ii) *be satisfied. If* $a(t) \not\equiv a(0)$ *and if* $u(t)$ *is any solution of* (31.20) *which exists on* $0 \leq t < \infty$, *then* $\lim_{t \to \infty} d^j u / dt^j = 0$, $j = 0, 1, 2$.

If the complete monotonicity of a in (i) above is replaced by

(i′) $a(t) \in C[0, \infty)$, $(-1)^k a^{(k)}(t) \geq 0$ $(0 < t < \infty; \ k = 0, 1, 2, 3)$,

it is shown in [111d]:

THEOREM 31.6. *Let* (i), (ii) *be satisfied. If* $a(t) \not\equiv a(0)$, *then the conclusion of Theorem* 31.5 *holds.*

The proof of Theorem 31.6 is based on the Lyapunov functional

$$E(t) = G(u(t)) + \frac{1}{2} a(t) \left(\int_0^t g(u(\tau)) \, d\tau \right)^2 - \frac{1}{2} \int_0^t a'(t - \tau) \left(\int_\tau^t g(u(s)) \, ds \right)^2 d\tau \, .$$

While the interest here is in the behavior of solutions of (31.20) as $t \to \infty$, it may be noted that the *a priori* bounds obtained in the proofs of Theorems 31.5 and 31.6, as well as in Theorem 31.7 below, and recent results in [141b] readily imply that every local solution of (31.20) may be continued (not necessarily uniquely) over $0 \leq t < \infty$. It may also be remarked that the proofs of Theorems 31.5 and 31.6 utilize not only $V(t)$, $E(t) \geq 0$; $V'(t)$, $E'(t) \leq 0$, but also boundedness of $|V''(t)|$, $|E''(t)|$ over $0 \leq t < \infty$, which, of course, imply $V'(t)$, $E'(t) \to 0$ as $t \to \infty$.

In [111e] the nonlinear delay equation

(31.21) $$\frac{dx}{dt} = -\int_{t-L}^t a(t - \tau) g(x(\tau)) \, d\tau \qquad (0 \leq t < \infty)$$

is considered with solutions satisfying the initial condition

(31.22) $$x(t) = \Psi(t) \qquad (-L \leq t \leq 0) \, ,$$

where $L > 0$ is a constant and where Ψ is a given real continuous function. Let $a(t) \in C[0, L]$ satisfy (i′) on $0 < t \leq L$, and let $a(L) = 0$; let $g(x)$ satisfy (ii). The method of Theorem 31.6, applied to the functional

$$G(u(t)) - \frac{1}{2} \int_{t-L}^t a'(t - \tau) \left(\int_\tau^t g(u(s)) \, ds \right)^2 d\tau \geq 0 \, ,$$

shows that if $a''(t) \not\equiv 0, 0 \leq t < L$ and if $u(t)$ is any solution of (31.21), (31.22) existing on $0 \leq t < \infty$, then the conclusion of Theorems 31.5 and 31.6 again holds. However, if $a''(t) \equiv 0$ and the other conditions are as above, this method gives only boundedness of $|d^k u/dt^k|$ on $0 \leq t < \infty$, $k = 0, 1, 2$. Therefore the equation

$$(31.23) \qquad \frac{dx}{dt} = -\frac{1}{L} \int_{t-L}^{t} (L - (t - \tau)) g(x(\tau))\, d\tau \quad (0 \leq t < \infty)\,,$$

with solutions satisfying (31.22), requires a more detailed study. Note that the kernel in (31.23) obviously satisfies the requirement of $a(t)$ in (31.21) above.

Let $g(x)$ satisfy (ii) as well as

(iii) for each $A > 0$ there exists a constant $K = K(A) > 0$, such that $|g(x) - g(y)| \leq K|x - y|$, where $|x|, |y| \leq A$.

If $(\alpha, \beta) \neq (0, 0)$, let $\Gamma = \Gamma(\alpha, \beta)$ be the (necessarily unique) periodic orbit of the ordinary differential equation

$$(31.24) \qquad \frac{d^2 x}{dt^2} + g(x(t)) = 0\,,$$

passing through the point (α, β) in the phase plane associated with (31.24). If $(\alpha, \beta) = (0, 0)$, let Γ be the origin.

THEOREM 31.7. *Let* (ii), (iii) *be satisfied. Let* $u(t)$ *be any solution of* (31.23) *which exists for* $t > 0$ *and satisfies* (31.22). *Then there exists a constant* $K > 0$ *such that* $|d^k u/dt^k| \leq K$, $0 \leq t < \infty$, $k = 0, 1, 2$. *Moreover, if* Ω *is the set of points* $\Omega = [(x, y); x = \lim_{n\to\infty} u(t_n), y = \lim_{n\to\infty} u'(t_n)$ *for some* $\{t_n\} \to \infty]$, *then there exists a unique orbit* Γ *of* (31.24) *such that* $\Omega = \Gamma$, *where* Γ *depends only on* Ψ. *If* $\Gamma \neq (0, 0)$, *then there exists an integer* $m \geq 1$ *such that* $L = m\gamma(\alpha, \beta)$, *where* $\gamma(\alpha, \beta)$ *is the least period of the solutions of* (31.24) *associated with the orbit* Γ *through the point* (α, β).

The proof of Theorem 31.7 makes use of the Lyapunov functional

$$G(u(t)) + \frac{1}{2L} \int_{t-L}^{t} \left(\int_{\tau}^{t} g(u(s))\, ds \right)^2 d\tau \geq 0\,,$$

and the periodicity of all solutions of (31.24), combined with the type of reasoning reminiscent of the study of autonomous systems of ordinary differential equations in the plane. Theorem 31.7 has been obtained by Brownell and Ergen [26a] but under considerably more severe hypotheses than (ii), (iii) and by different (less elementary) methods.

7

Application of Lyapunov's Method to the Problem of Stability of Equations with Time Delay

In this chapter we consider applications of the theorems given in the preceding chapter to certain problems of stability for equations with time delay. In section 32 we consider the question of stability for a persistent disturbance. In section 33 we consider first-order stability. In section 34 we construct specific functionals for certain systems with time delay.

32. Stability for Persistent Disturbances

The general theorems proved in the preceding sections can be extended to equations with time delay. The methods used are in essence an extension of Lyapunov's second method for ordinary equations. In this section we consider in particular the problem of stability for a persistent disturbance, when the equation involves delay in the argument t. We first make a definition that corresponds to the definition previously made for ordinary equations when there is a persistent disturbance. (See section 24, p. 101.)

Let us consider the equation with time delay

$$(32.1) \qquad \frac{dx_i}{dt} = X_i\Big(x_1(t), \cdots, x_n(t); \ x_1(t - h_{i1}(t)), \cdots, x_n(t - h_{in}(t)), t \Big) ,$$

$$(0 \leq h_{ij}(t) \leq h; \ i, j = 1, \cdots, n) ,$$

where the continuous functions $X_i(x_1, \cdots, x_n; y_1, \cdots, y_n, t)$ are defined for $|x_j| < H, |y_j| < H$ and satisfy a Lipschitz condition with respect to x_j, y_j; i.e.,

$$(32.2) \qquad | X_i(x_1'', \cdots, x_n''; y_1'', \cdots, y_n'', t) - X_i(x_1', \cdots, x_n'; y_1', \cdots, y_n', t) |$$

$$\leq L(| x_1'' - x_1' | + \cdots + | x_n'' - x_n' | + | y_1'' - y_1' | + \cdots + | y_n'' - y_n' |) .$$

Moreover, we assume as usual that $X_i(0, \cdots, 0; 0, \cdots, 0, t) = 0$ $(i = 1, \cdots, n)$.

In addition to the system (32.1), we consider the "perturbed" system of equations

$$(32.3) \qquad \frac{dx_i}{dt} = X_i\Big(x_1(t), \cdots, x_n(t); x_1(t - h_{i1}^*(t)), \cdots, x_n(t - h_{in}^*(t)), t \Big)$$

$$+ R_i\Big(x_1(t), \cdots, x_n(t); x_1(t - g_{ij}(t)), \cdots, x_n(t - g_{ij}(t)), t \Big)$$

$$(i, j = 1, \cdots, n; 0 \leq h_{ij}^*(t) \leq h; 0 \leq g_{ij}(t) \leq h) ,$$

where the continuous functions R_i are not required to reduce to 0 for $x_j = y_j = 0$.

DEFINITION 32.1.   *The null solution $x = 0$ of the system (32.1) is called stable for persistent disturbances if for every $\varepsilon > 0$ there exist positive numbers δ_0, η, Δ such that the solution $x(x_0(\vartheta_0), t_0, t)$ of the system (32.3) satisfies the inequality*

$$(32.4) \qquad\qquad || x(x_0(\vartheta_0), t_0, t) || < \varepsilon$$

for all $t \geqq t_0, t_0 \geqq 0$, whenever the initial curve $x_0(\vartheta_0)$ satisfies the inequality

$$(32.5) \qquad\qquad || x_0(\vartheta_0) ||^{(h)} \leqq \delta_0$$

and the perturbed time delays h_{ij}^* and the functions R_i satisfy the inequalities

$$(32.6) \qquad | R_i(x_1, \cdots, x_n; y_1, \cdots, y_n, t) | < \eta \qquad (i = 1, \cdots, n)$$

for $|| x || < \varepsilon, || y || < \varepsilon$, and

$$(32.7) \qquad\qquad | h_{ij}(t) - h_{ij}^*(t) | < \Delta \qquad (i, j = 1, \cdots, n) .$$

In section 24 (p. 102) we saw that for ordinary equations a sufficient condition for stability under persistent disturbances is asymptotic stability that is uniform with respect to the coordinates and the time of the initial perturbation; the proof of this fact is based on the theorem on the existence of a Lyapunov function. The corresponding theorem needed in the present context is theorem 30.2, which asserts the existence of the functional V. The analogous facts for equation (32.1) with time delay are given by the following theorem.

THEOREM 32.1.   *Suppose that the null solution $x = 0$ of equation (32.1) is asymptotically stable uniformly with respect to time t_0 and the initial curve $x_0(\vartheta_0)$, in the sense of definition 30.2 (p. 146). Then the null solution $x = 0$ is stable for persistent disturbances.*

Remark.   If X_i and $h_{ij}(t)$ are periodic functions of the time t of period θ, or are independent of the time, then (as was shown on p. 150), the asymptotic stability of the null solution $x = 0$ is always uniform with respect to $x_0(\vartheta_0)$ and t_0. It follows in this case that the null solution $x = 0$ is stable under persistent disturbances when it is asymptotically stable.

PROOF OF THE THEOREM.   By theorem 30.2, the hypotheses of theorem 31.1 guarantee the existence of a functional V that satisfies the conditions (30.1) and moreover satisfies a Lipschitz condition (30.15) with respect to $x(\vartheta)$. Now let ε $(\varepsilon > 0)$ be an arbitrary positive number. Since the functional $V(x(\vartheta), t)$ is positive-definite, there is a constant $\beta > 0$ with the property

$$(32.8) \qquad v(x(\vartheta), t) > \beta \quad \text{for } || x(\vartheta) ||^{(h)} = \varepsilon, \ t \geqq 0 .$$

Since the functional V admits an infinitely small upper bound, there must be a number $\delta > 0$ with the property

(32.9) $$V(x(\vartheta), t) < \beta \quad \text{for } \| x(\vartheta) \|^{(h)} = \delta, \ t \geqq 0 .$$

In the region

(32.10) $$\delta \leqq \| x(\vartheta) \|^{(h)} \leqq \varepsilon$$

the inequality

(32.11) $$\lim_{\Delta t \to +0} \sup \left(\frac{\Delta V}{\Delta t} \right)_{(32.1)} \leqq -\alpha$$

is satisfied for some positive constant $\alpha > 0$. This is the meaning of negative-definiteness. Now we calculate the value of the quantity

(32.12) $$\lim_{\Delta t \to +0} \sup \left(\frac{\Delta V}{\Delta t} \right)_{(32.3)}$$

along trajectories of the system (32.3):

$$\lim_{\Delta t \to +0} \sup \left(\frac{\Delta V}{\Delta t} \right)_{(32.3)} \leqq \lim_{\Delta t \to +0} \sup \left(\frac{\Delta V}{\Delta t} \right)_{(32.1)}$$
$$+ \lim_{\Delta t \to +0} \sup \left(\frac{1}{\Delta t} [V(x^{(1)}(t + \Delta t + \vartheta), t + \Delta t) \right.$$
$$\left. - V(x^{(3)}(t + \Delta t + \vartheta), t + \Delta t)] \right) ,$$

where the symbols $x^{(1)}(t)$, $x^{(3)}(t)$ denote trajectories of systems (32.1) and (32.3), respectively. By condition (30.15), we have the inequalities

$$| V(x^{(1)}(t + \Delta t + \vartheta), t + \Delta t) - V(x^{(3)}(t + \Delta t + \vartheta), t + \Delta t) |$$
$$< K \| x^{(1)}(t + \Delta t + \vartheta) - x^{(3)}(t + \Delta t + \vartheta) \|^{(h)} .$$

On the other hand, applying inequalities (32.2) to equations (32.1) and (32.3), we see that

$$\lim_{\Delta t \to +0} \sup \left(\frac{1}{\Delta t} [\| x^{(1)}(t + \Delta t + \vartheta) - x^{(3)}(t + \Delta t + \vartheta) \|^{(h)}] \right)$$
$$\leqq \sup | R_i | + L \left(\sum_{i-1}^{n} | x_j(t - h_{ij}^*) - x_j(t - h_{ij}) | \right) .$$

We turn to the summands in the last expression. Let $t > 2h$. Suppose that for $\tau < t$, the trajectory $x^{(3)}(\tau)$ lies in the region $\| x(\vartheta) \|^{(h)} < \varepsilon$. Then the functions $x_j^{(3)}(\tau)$ satisfy a Lipschitz condition with respect to τ for some constant L_1 (section 28, p. 133):

$$| x_j(t - h_{ij}) - x_j(t - h_{ij}^*) | \leqq | h_{ij} - h_{ij}^* | L_1 .$$

With this inequality in hand, we obtain the further estimate

$$\lim_{\Delta t \to +0} \sup \left(\frac{\Delta V}{\Delta t} \right)_{(32.3)} \leqq -\alpha + K(\sup | R_i | + n^2 L L_1 \sup | h_{ij} - h_{ij}^* |) ;$$

this is valid along all the integral curves of (32.3) in the region (32.10) for $t > t_0 + 2h$ whenever these trajectories $x_i(\tau)$ of (32.3) lie in the region

$||x|| < \varepsilon$ for all $\tau < t$. If the functions R_i and h_{ij}^* satisfy inequalities (32.6) and (32.7), where $\eta < \alpha/(4K)$ and $\varDelta < \alpha/(4n^2 LL_1)$, then the estimate

$$(32.13) \qquad \limsup_{\varDelta t \to +0} \left(\frac{\varDelta V}{\varDelta t} \right)_{(32.3)} < -\frac{\alpha}{2}$$

is valid in the region (32.10) along all the integral curves of (32.3) satisfying the above hypothesis. In view of conditions (32.8), (32.9), and (32.13), we conclude that when t increases further, the trajectories in question do not leave the region $||x|| < \varepsilon$. Now let us apply (27.12). Whenever $||x_0(\vartheta_0)|| < \delta e^{-2hL}$, it follows that the conditions postulated above are valid for $t = t_0 + 2h$ along an arbitrary trajectory of (32.3). Thus we have shown that the inequality $||x(x_0(\vartheta_0), t_0, t)|| < \varepsilon$ is valid, provided that inequalities (32.6) and (32.7) can be satisfied with positive constants η and $\varDelta$ that are sufficiently small, and provided further that $||x_0(\vartheta_0)||^{(h)} < \delta e^{-2hL}$. Thus the choice $\delta_0 = \delta e^{-2hL}$ does satisfy all the conditions needed. The theorem is proved.[*]

Remark. If a persistent disturbance is bounded in the mean, the result established for ordinary equations (section 24, p. 102) can be extended to equations with time delay. To obtain this extension, one must replace the function $f = v(x, t)e^{\beta(t)}$ used there by the functional $F = V(v(\vartheta), t)e^{\beta(t)}$. The result in question was formulated by Germaĭdze. We content ourselves with giving the result.

Let T be a fixed positive constant $(T > 0)$. Consider various functions R_i and $h_{ij}^*(t)$ and suppose there exist corresponding functions $\psi_1(t)$ and $\psi_2(t)$ such that the inequalities

$$(32.14) \qquad |R_i(x_1, \cdots, x_n; y_1, \cdots, y_n, t)| < \psi_1(t)$$

$$\text{for } |x_i| < \varepsilon, \ |y_i| < \varepsilon, \ \sup |h_{ij}(t) - h_{ij}^*(t)| < \psi_2(t),$$

$$(32.15) \qquad \int_{t_0}^{t_0+T} \psi_1(\xi)\, d\xi < \eta, \qquad \int_{t_0}^{t_0+T} \psi_2(\xi)\, d\xi < \varDelta,$$

hold.

If the null solution $x = 0$ of equation (32.1) is asymptotically stable, uniformly in $x_0(\vartheta_0)$ and t_0, then for every positive number $\varepsilon > 0$, however small, we can find numbers $\eta > 0$, $\delta_0 > 0$, and $\varDelta > 0$ such that whenever relations (32.14) and (32.15) hold, and the initial value $x_0(\vartheta_0)$ satisfies inequality (32.5), the solution $x(x_0(\vartheta_0), t_0, t)$ of the system of equations (32.3) will satisfy the inequality $||x(x_0(\vartheta_0), t_0, t)|| < \varepsilon$ for all $t \geqq t_0$, $t_0 \geqq 0$.

33. First-order Stability for Equations with Time Delay

Virtually the entire theory of first-order stability for ordinary differential equations carries over to equations with time delay.

An outline of the facts will suffice. First we generalize theorem 11.1, which asserts the existence of the function $v(x, t)$. The corresponding

---

[*] Repin [163] solved the first-order stability problem for equations with time delay and also the problem of stability with persistent disturbances.

theorem on the existence of a functional $V(x(\vartheta), t)$ is theorem 33.1, proved below.   This theorem applies to systems of equations

$$(33.1) \qquad \frac{dx_i}{dt} = X_i(x_1(t + \vartheta), \cdots, x_n(t + \vartheta), t)$$

with time delay, where the functionals X_i satisfy the conditions imposed in section 27 (pp. 127-28).   We first prove a lemma.

LEMMA 33.1.   *If the solution* $x(x_0(\vartheta_0), t_0, t)$ *of system* (33.1) *satisfies the condition*

$$(33.2) \qquad || x(x_0(\vartheta_0), t_0, t) ||^{(h)} \leqq B || x_0(\vartheta_0) ||^{(h)} \exp(-\alpha(t - t_0)), \qquad t \geqq t_0 ,$$

in the region

$$(33.3) \qquad\qquad || x_0(\vartheta_0) ||_h < H_0 = \frac{H}{B} ,$$

then we can find a functional $V(x(\vartheta), t)$ *that satisfies the following conditions in the region* (33.3):

$$(33.4) \qquad\qquad c_1 || x(\vartheta) ||^{(h)} \leqq V(x(\vartheta), t) \leqq c_2 || x(\vartheta) ||^{(h)} ,$$

$$(33.5) \qquad\qquad \limsup_{\Delta t \to +0} \left(\frac{\Delta V}{\Delta t} \right)_{(33.1)} \leqq -c_3 || x(\vartheta) ||^{(h)} ,$$

$$(33.6) \qquad | V(x''(\vartheta), t) - V(x'(\vartheta), t) | \leqq c_4 || x''(\vartheta) - x'(\vartheta) ||^{(h)}$$

($c_1, \cdots, c_4$ *are positive constants*).

PROOF.   We define the functional

$$(33.7) \qquad V(x_0(\vartheta_0), t_0) = \int_{t_0}^{t_0+T} || x(x_0(\vartheta_0), t_0, \xi + \vartheta) ||^{(h)} d\xi$$
$$+ \sup [|| x(x_0(\vartheta_0), t_0, \xi + \vartheta) ||^{(h)} \quad \text{for} \quad t_0 \leqq \xi \leqq t_0 + T]$$
$$\left(T = \frac{1}{\alpha} \ln 2B \right)$$

and show that it has the stated properties.   In the first place, relation (33.2) shows that if the initial curve $x_0(\vartheta_0)$ lies in the region (33.3), the solution $x(x_0(\vartheta_0), t_0, t)$ lies in the region $|| x(\vartheta) ||^{(h)} < H$ for $t \geqq t_0$.   For the right members of system (33.1) are defined in this region, and therefore the solution $x(x_0(\vartheta_0), t_0, t)$ can be extended for all $t \geqq t_0$.   Thus the functional (33.7) is defined in the region (33.3).   The left inequality of (33.4) (even with $c_1 = 1$) follows from the obvious relations

$$\sup [|| x(x_0(\vartheta_0), t_0, \xi + \vartheta) ||^{(h)} \quad \text{for} \quad t_0 \leqq \xi \leqq t_0 + T] \geqq || x_0(\vartheta_0) ||^{(h)} .$$

The right inequality of (33.4) follows from condition (33.2) and the relation

$$V(x_0(\vartheta_0), t_0) \leqq \int_{t_0}^{t_0+T} B || x_0(\vartheta_0) ||^{(h)} \exp[-\alpha(t - t_0)] \, dt + || x_0(\vartheta_0) ||^{(h)} B$$
$$= B\left[1 + \frac{1}{\alpha}\left(1 - \frac{1}{2B} \right) \right] || x_0(\vartheta_0) ||^{(h)} .$$

Relation (33.2) shows that the second term of the right member of equation (33.7) does not increase with increasing t_0 [along trajectories of the system (33.1)]. Thus

$$\limsup_{\substack{\Delta t \to +0 \\ \text{for } t=t_0 \\ \text{along (33.1)}}} \left(\frac{\Delta V}{\Delta t} \right)$$

$$\leq \lim_{\Delta t \to +0} \left(\int_{t_0+\Delta t}^{t_0+T+\Delta t} || x(x_0(\vartheta_0), t_0, \xi + \vartheta) ||^{(h)} d\xi - \int_{t_0}^{t_0+T} || x(x_0(\vartheta_0), t_0, \xi + \vartheta) ||^{(h)} d\xi \right)$$

$$= -|| x_0(\vartheta_0) ||^{(h)} + || x(x_0(\vartheta_0), t_0, t_0 + T + \vartheta) ||^{(h)} .$$

Therefore, relation (33.2) and the choice of the number T ensure that

$$\limsup_{\substack{\Delta t \to +0 \\ \text{for } t=t_0 \\ \text{along (33.1)}}} \left(\frac{\Delta V}{\Delta t} \right) \leq -\tfrac{1}{2} || x_0(\vartheta_0) ||^{(h)} ,$$

and this demonstrates inequality (33.5) (even with c_3 taken as $\tfrac{1}{2}$).

Now we come to inequality (33.6). We use (33.2) and substitute into the integral to obtain

$$| V(x_0''(\vartheta_0), t_0) - V(x_0'(\vartheta_0), t_0) |$$

$$\leq \int_{t_0}^{t_0+T} || x(x_0''(\vartheta_0), t_0, \xi + \vartheta) - x(x_0'(\vartheta_0), t_0, \xi + \vartheta) || d\xi$$

$$+ \sup \left(|| x(x_0''(\vartheta_0), t_0, \xi + \vartheta) - x(x_0'(\vartheta_0), t_0, \xi + \vartheta) ||^{(h)} \right.$$

$$\left. \text{for } t_0 \leq \xi \leq t_0 + T \right)$$

$$\leq \int_{t_0}^{t_0+T} || x_0''(\vartheta_0) - x_0'(\vartheta_0) ||^{(h)} e^{nL\xi} d\xi + || x_0''(\vartheta_0) + x_0'(\vartheta_0) ||^{(h)} e^{nLT}$$

$$= c_4 || x_0'(\vartheta_0) - x_0''(\vartheta_0) ||^{(h)} .$$

This establishes inequality (33.6), and the lemma is proved.

The following theorem is a consequence of theorem 29.1 and lemma 33.1.

THEOREM 33.1. *Let* $X_i(x_1(\vartheta), \cdots, x_n(\vartheta))$ *be linear functionals. If the roots* λ *of the characteristic equation*

$$(33.8) \qquad \det \begin{bmatrix} X_{11} - \lambda \cdots & X_{1n} \\ \cdots\cdots\cdots\cdots\cdots \\ X_{n1} & \cdots X_{nn} - \lambda \end{bmatrix} = 0 \qquad (X_{ij} = X_i(0, \cdots, e_j^{\lambda\vartheta}, \cdots, 0))$$

satisfy the inequality

$$(33.9) \qquad\qquad\qquad \operatorname{Re} \lambda < -\gamma$$

for some positive constant $\gamma > 0$, *then there is a functional* $V(x(\vartheta), t)$ *that satisfies conditions* (33.4)–(33.6).

Remark. As the proof of lemma 33.1 shows, the existence of a functional $V(x(\vartheta), t)$ for which the estimates (33.4)–(33.6) are valid does not depend primarily on the fact that the functionals X_i are linear. It depends rather on the specific type of estimate (33.2) that is valid along the trajectories $x(x_0(\vartheta_0), t_0, t)$ leaving the point $x = 0$ as t increases. It also follows that inequality (33.2) is characteristic of linear systems whose right members are independent of the time t.

From theorem 33.1 we can obtain first-order stability theorems for equations with time delay. We now prove one such theorem.

We consider the system of equations

$$(33.10) \qquad \frac{dx_i}{dt} = X_i(x_1(\vartheta), \cdots, x_n(\vartheta)) + R_i(x_1(\vartheta), \cdots, x_n(\vartheta), t) \,,$$

where X_i are linear functionals and R_i are arbitrary continuous functionals that we assume satisfy the inequality

$$(33.11) \qquad | R_i(x_1(\vartheta), \cdots, x_n(\vartheta), t) | \leq \beta \, || \, x(\vartheta) \, ||^{(h)} \,.$$

THEOREM 33.2. *If the roots λ of equation (33.8) satisfy inequality (33.9), then we find a positive constant $\beta > 0$ such that the null solution $x = 0$ of system (33.10) is asymptotically stable for arbitrarily chosen continuous functionals R_i subject to condition (33.11).*

PROOF. Theorem 33.1 shows that the hypotheses of theorem 33.2 guarantee the existence of a functional $V(x(\vartheta), t)$ that satisfies conditions (33.4)–(33.6). Evaluation of the functional in question along trajectories of the system (33.10) proceeds as follows:

$$\lim_{\Delta t \to +0} \sup \left(\frac{\Delta V}{\Delta t} \right)_{(33.10)}$$

$$= \lim_{\Delta t \to +0} \sup \left(\frac{1}{\Delta t} [V(x^{(10)}(t + \Delta t + \vartheta), t + \Delta t) - V(x^{(10)}(t + \vartheta), t)] \right)$$

$$\leq \lim_{\Delta t \to +0} \sup \left(\frac{1}{\Delta t} [V(x^{(1)}(t + \Delta t + \vartheta), t + \Delta t) - V(x^{(1)}(t + \vartheta), t)] \right)$$

$$+ \lim_{\Delta t \to +0} \sup \left(\frac{1}{\Delta t} [V(x^{(10)}(t + \Delta t + \vartheta), t + \Delta t) - V(x^{(1)}(t + \Delta t + \vartheta), t + \Delta t)] \right)$$

$$\leq -c_3 \, || \, x(t + \vartheta) \, ||^{(h)}$$

$$+ c_4 \lim_{\Delta t \to +0} \sup \left(\frac{1}{\Delta t} [|| \, x^{(10)}(t + \Delta t + \vartheta) - x^{(1)}(t + \Delta t + \vartheta) \, ||^{(h)}] \right) .$$

Here the symbols $x^{(10)}(t)$ and $x^{(1)}(t)$ denote trajectories of the systems (33.10) and (33.1), respectively. Using inequality (33.11) and equation (33.10), we obtain the following estimate:

$$(33.12) \qquad \lim_{\Delta t \to +0} \sup \left(\frac{\Delta V}{\Delta t} \right)_{(33.10)} \leq (-c_3 + c_4\beta) \, || \, x(t + \vartheta) \, ||^{(h)} \,.$$

If the positive number β ($\beta > 0$) satisfies the condition

(33.13) $$\beta > \frac{c_3}{c_4} > 0 \, ,$$

then the value of $\lim_{\Delta t \to +0} \sup (\Delta V/\Delta t)_{(33.10)}$ is a negative-definite functional. Thus the functional $Vx((\vartheta), t)$ does satisfy the conditions of theorem 33.1 along trajectories of the system (33.10), and the null solution $x = 0$ is therefore asymptotically stable. The theorem is proved.

In the special equation with time delay that has the form

(33.14) $$\frac{dx_i}{dt} = \sum_{j=1}^{n} [a_{ij} x_j(t) + b_{ij} x_j(t - h_{ij})]$$
$$+ R_i(x_1, \cdots, x_n, x_1(t - h_{ij}^*(t)), \cdots, x(t - h_{ij}^*(t)), t)$$

with h_{ij} constant, $0 \leq h_{ij} \leq h$, and $0 \leq h_{ij}^*(t) \leq h$, theorem 33.2 takes the following form:

If the roots of the characteristic equation

(33.15) $$\det \begin{bmatrix} a_{11} \exp(-\lambda h_{11}) - \lambda \cdots & a_{1n} \exp(-\lambda h_{1n}) \\ \cdots\cdots\cdots\cdots\cdots\cdots\cdots\cdots\cdots\cdots\cdots \\ a_{n1} \exp(-\lambda h_{n1}) & \cdots & a_{nn} \exp(-\lambda h_{nn}) - \lambda \end{bmatrix} = 0$$

satisfy the inequality

(33.16) $$\operatorname{Re} \lambda < -\gamma$$

for some positive constant $\gamma > 0$, then we can find a positive constant $\beta > 0$ such that the null solution $x = 0$ of system (33.14) is asymptotically stable for every continuous function $R_i(x_1, \cdots, x_n; y_1, \cdots, y_n, t)$ and every delay $h_{ij}^*(t)$ that satisfies the inequalities

(33.17) $$| R_i(x_1, \cdots, x_n; y_1, \cdots, y_n, t) | \leq \beta(\| x \| + \| y \|) \, .$$

Remark. This theorem includes as a special case certain results of Bellman [19], [20] and Wright [188] on first-order stability of equations with time delay. Furthermore, the results we have just obtained can be generalized to the case in which the inequalities of the form (33.11) are postulated to hold for the mean values of the functions R_i for a certain interval T of the time. The method of proof of this latter assertion corresponds to the method used for ordinary equations on pp. 102–5, except that in the place of the function $v(x, t)$, we use the functional $V(x(\vartheta), t)$. The result is as follows.[*]

If the roots λ of equation (33.8) satisfy inequality (33.9), then we can find a positive constant $\beta > 0$ such that the null solution $x = 0$ of the system (33.10) is asymptotically stable for arbitrary choice of initial functionals $R_i(x(\vartheta), t)$ satisfying the inequality

$$| R_i(x(\vartheta), t) | \leq \psi(t) \| x(\vartheta) \|^{(h)} \, ,$$

where

---

[*] The proof is due originally to Germaĭdze.

(33.18)
$$\frac{1}{T}\int_t^{t+T}\psi(\xi)d\xi < \beta\ .$$

If in particular the delays are functions $h_{ij}(t)$, the assertion specializes as follows:

If the roots λ of equation (33.15) satisfy inequality (33.16), then there is a positive constant $\beta > 0$ such that the null solution $x = 0$ of system (33.14) is asymptotically stable for arbitrary choice of the delays $h_{ij}^*(t)$, and continuous functions $R_i(x, y, t)$ satisfying the inequalities

$$|\,R_i\,(x, y, t)\,| \leqq \psi(t)(||\,x\,|| + ||\,y\,||)$$

where the function $\psi(t)$ satisfies inequality (33.18).

34. Construction of Functionals in Special Cases

We begin our consideration of special examples of functionals with the system of equations with time delay

(34.1)
$$\frac{dx_i}{dt} = a_{i1}x_1(t) + \cdots + a_{in}x_n(t)$$
$$+ \varphi_i(x_1(t), \cdots, x_n(t);\ x_1(t - h_{i1}), \cdots, x_n(t - h_{in}), t)$$
$$(i = 1, \cdots, n;\ a_{ij} = \text{const.};\ h_{ij} = \text{const.})\ ,$$

where the functions φ_i satisfy the relation

(34.2)
$$|\,\varphi_i(x_1, \cdots, x_n;\ y_1, \cdots, y_n,\ t)\,| \leqq \beta(||\,x\,|| + ||\,y\,||)\ .$$

We assume that the roots λ of the characteristic equation

(34.3)
$$\det\begin{bmatrix} a_{11} - \lambda \cdots a_{1n} \\ a_{21} \quad \cdots \quad a_{2n} \\ \cdots\cdots\cdots\cdots \\ a_{n1} \quad \cdots \ a_{nn} - \lambda \end{bmatrix} = 0$$

have negative real part. In this case, Lyapunov [114, pp. 276–77] showed that the quadratic form

(34.4)
$$v(x_1, \cdots, x_n) = \sum_{i,j=1}^{n} b_{ij}x_ix_j$$

satisfies the condition

(34.5)
$$\sum_{i=1}^{n}\frac{\partial v}{\partial x_i}\left(\sum_{j=1}^{n}a_{ij}x_j\right) = -\sum_{i=1}^{n}x_i^2\ .$$

In other words, the condition

(34.6)
$$\sum_{j=1}^{n}(b_{ij}a_{jk} + b_{kj}a_{ji}) = -\delta_{ik}\begin{cases}\delta_{ii} = 1, \\ \delta_{ik} = 0 \ \text{for}\ i \neq k\end{cases}$$

is valid.

In this case the functional $V(x(\vartheta), t)$ can be taken in the form

$$(34.7) \qquad V(x(\vartheta)) = v(x_1, \cdots, x_n) + \sum_{i,j=1}^{n} \varepsilon_{ij} \int_{-h_{ij}}^{0} x_j^2(\vartheta) d\vartheta .$$

The value of the derivative dV/dt along a trajectory of the system (34.1) is calculated as follows:

$$(34.8) \quad \lim_{\Delta t \to +0} \left(\frac{\Delta V}{\Delta t} \right)_{(34.1)} = \sum_{i=1}^{n} \frac{\partial v}{\partial x_i} \sum_{j=1}^{n} a_{ij} x_j$$

$$+ \sum_{i=1}^{n} \frac{\partial v}{\partial x_i} \varphi_i(x_1, \cdots, x_n; \ x_1(t - h_{i1}), \cdots, x_n(t - h_{in}), t)$$

$$+ \sum_{i,j=1}^{n} \varepsilon_{ij} x_j^2 - \sum_{i,j=1}^{n} \varepsilon_{ij} x_j^2(t - h_{ij}) .$$

Functions of the arguments $x_i = x_i(t)$ and $y_{ij} = x_i(t - h_{ij})$ appear in the right member of equation (34.8). If we can choose the constants $\varepsilon_{ij} \geq 0$ that appear in the right member of equation (34.8) as nonnegative and such that this right member is negative-definite, then the null solution $x = 0$ of system (34.1) is asymptotically stable. Indeed, in this case all the conditions of theorem 31.3 are satisfied, since by Lyapunov's theorem, the form $v(x_1, \cdots, x_n)$ is positive-definite. Thus there is a positive constant $\gamma > 0$ that satisfies the inequality

$$V(x(\vartheta), t) \geq v(x_1, \cdots, x_n) \geq \gamma \| x \|^2 \quad (\gamma > 0) .$$

But if $\beta > 0$ is positive and sufficiently small, it is always possible to select the numbers $\varepsilon_{ij} \geq 0$ so that the right member of equation (34.8) is a negative-definite function. For example, if we take $\varepsilon_{ij} = 1/(2n^2)$, the estimates

$$\lim_{\Delta t \to +0} \left(\frac{\Delta V}{\Delta t} \right)_{(34.1)} \leq - \sum_{i=1}^{n} x_i^2 + \frac{1}{2} \sum_{i,j=1}^{n} x_i^2(t - h_{ij}) + \left| \left(\sum_{i=1}^{n} a_{ij} x_j \right) \beta \sum_{i,j=1}^{n} | x_i(t - h_{ij}) | \right|$$

$$\leq - \frac{1}{2} \sum_{i=1}^{n} x_i^2(t) - \sum_{i,j=1}^{n} \frac{1}{2n^2} x_i^2(t - h_{ij}) \beta N \sum_{i,j=1}^{n} (x_i^2 + x_i^2(t - h_{ij}))$$

$$(N = \text{const.})$$

are valid.

For a particular case, we consider the first-order differential equation

$$(34.9) \qquad \frac{dx}{dt} = -ax + b(t)x(t - h) ,$$

where a is a positive constant, $a > 0$. By the general theory the functional $V(x(\vartheta))$ has the form

$$V(x(\vartheta)) = \frac{1}{2a} x^2 + \mu \int_{-h}^{0} x^2(\vartheta) d\vartheta .$$

The derivative dV/dt of this functional along a trajectory of the system (34.9) has the value

$$(34.10) \qquad \lim_{\Delta t \to +0} \left(\frac{\Delta V}{\Delta t} \right) = -x^2(t) + \frac{b(t)}{a} x(t)x(t - h) + \mu x^2(t) - \mu x^2(t - h) .$$

To make the right member of (34.10) a negative-definite function of the arguments $x(t)$ and $x(t - h)$, it is sufficient to satisfy the inequality

$$(34.11) \qquad 4(1 - \mu)\mu > \frac{b^2(t)}{a^2} .$$

Thus the null solution $x = 0$ of equation (34.9) is certainly asymptotically stable if there is a positive constant $\mu > 0$ that satisfies relation (34.11). Now the left member of (34.11) reaches its maximum value for $\mu = \frac{1}{2}$, and in this case, relation (34.11) reads

$$b^2(t) < a^2 , \text{ or } | b(t) | < a .$$

Remark. The above method of constructing a functional $V(x(\vartheta))$ also applies if the functions $h_{ij}(t)$ are not constants. In this case, the value of the derivative of the functional V along a trajectory of (34.1) involves terms $\varepsilon_{ij}x_i^2(t - h_{ij})$, multiplied respectively by $(1 - h'_{ij}(t))$. The question of negative-definiteness of the function dV/dt involves estimations of the corresponding terms.

If the functions $a_{ij}(t)$ are also variable, the above methods apply and are subject to convenient modification involving in some cases multiplication of the functional V by a suitably chosen function $\psi(t)$. Sometimes only a portion of the functional V is multiplied by a suitable function $\psi(t)$. As an example, we consider the equation

$$\frac{dx}{dt} = -a(t)x + b(t)x(t - h(t)) ,$$

and construct the functional

$$(34.12) \qquad V(x(\vartheta), t) = \psi(t)x^2 + \mu \int_{-h(t)}^{0} x^2(\vartheta) \, d\vartheta$$

$$(\lambda_1 \leqq \psi(t) \leqq \lambda_2 , \text{ with } \lambda_1, \lambda_2 \text{ positive constants}) .$$

The null solution $x = 0$ is asymptotically stable, provided that the function

$$\lim_{\Delta t \to +0} \left(\frac{\Delta V}{\Delta t} \right) = -2a(t)\psi(t)x^2(t) + 2b(t)\psi(t)x(t)x(t - h(t))$$

$$+ \psi'(t)x^2(t) + \mu x^2(t) - \mu(1 - h'(t))x^2(t - h(t))$$

is a negative-definite function of the arguments $x(t)$, $x(t - h(t))$. This condition amounts to the inequality

$$(2a(t)\psi(t) - \mu - \psi'(t))(1 - h'(t))\mu > b^2(t)\psi^2(t) .$$

Now we consider a case in which we construct a function $v(x, t)$ that satisfies the conditions of theorem 31.4. Let the h_{ij} be arbitrary functions of the time t, and let the system of equations being considered be system (34.1). We also assume that the roots of equation (34.3) have negative real part. The function $v(x_1, \cdots, x_n)$ can simply be taken to satisfy condition (34.5). The derivative dv/dt along a trajectory of the system (34.1) has the form

$$(34.13) \quad \frac{dv}{dt} = \sum_{i=1}^{n} \frac{\partial v}{\partial x_i} \left(\sum_{j=1}^{n} a_{ij} x_j \right)$$

$$+ \sum_{j=1}^{n} \frac{\partial v}{\partial x_i} \varphi_i(x_1, \cdots, x_n; x_1(t - h_{ij}), \cdots, x_n(t - h_{ij}), t) .$$

To show that the hypotheses of theorem 31.4 are satisfied, we must show that the derivative dv/dt is a negative-definite function on curves $x_i(\xi)$ ($\xi < 0$), which satisfy the relation

$$(34.14) \quad v(x_1(0), \cdots, x_n(0)) > qv(x_1(\xi), \cdots, x_n(\xi))$$

$$\text{for } \xi < 0, \ 0 < q < 1, \ q = \text{const.}$$

The curves $x_i(\xi)$ that satisfy relation (34.14) are necessarily contained in the larger class of curves $x_i(\xi)$ that satisfy the relation

$$(34.15) \quad \sum_{i=1}^{n} x_i^2(0) \geq q \frac{\rho_{\min}}{\rho_{\max}} \sum_{i=1}^{n} x_i^2(\xi) ,$$

where $\rho_{\min}$ and $\rho_{\max}$ are the roots of minimum and maximum modulus, respectively, of the equation

$$\det \begin{bmatrix} b_{11} - \rho & \cdots & b_{1n} \\ b_{21} & \cdots & b_{2n} \\ \cdots\cdots\cdots\cdots \\ b_{n1} & \cdots & b_{nn} - \rho \end{bmatrix} = 0 .$$

Thus the null solution $x = 0$ of system (34.1) is asymptotically stable if the right member of equation (34.13) is a negative-definite function on all curves $x_i(\xi)$ that satisfy relation (34.15). In particular, the equation

$$\frac{dx}{dt} = -ax(t) + b(t)x(t - h(t))$$

can be investigated by means of the function $v = x^2/(2a)$. Along trajectories, we have

$$\frac{dv}{dt} = -x^2(t) + \frac{b(t)}{a} x(t)x(t - h(t)) .$$

Therefore the null solution $x = 0$ of the equation in question is asymptotically stable if the derivative dv/dt is a negative-definite function on all curves satisfying $x^2(0) \geq qx^2(\xi)$, and a sufficient condition for this is the relation

$$|b(t)| < qa \qquad (0 < q < 1) .$$

Certain nonlinear equations are also easily investigated by use of Lyapunov functions.

As a first example, we consider the second-order nonlinear equation

(34.16)
$$\frac{d^2x}{dt^2} + \varphi\left(\frac{dx}{dt}, t\right) + f(x(t - h(t))) = 0 ,$$

where $f(x)$ is a continuously differentiable function satisfying the condition

(34.17)
$$\frac{f(x)}{x} > a > 0 , \quad |f'(x)| < L \quad \text{for } x \neq 0 ,$$

and $\varphi(y, t)$, $h(t)$ are continuous periodic functions of the time t satisfying the conditions

(34.18)
$$\frac{\varphi(y, t)}{y} > b > 0 \quad \text{for } y \neq 0 ,$$

and

(34.19)
$$h(t) \geq 0 , \quad h(t) \leq h \quad (h = \text{const.}) .$$

To take account of the meaning of equation (34.16) for negative values of t, it is necessary to write the equation so that it includes the last term (the integral) of the second equation of the following system:

(34.20)
$$\frac{dx}{dt} = y ,$$
$$\frac{dy}{dt} = -\varphi(y(t), t) - f(x(t)) + \int_{-h(t)}^{0} f'(x(t + \vartheta))y(t + \vartheta) \, d\vartheta .$$

This system is the exact meaning of equation (34.16). We consider the functional

(34.21)
$$V(x(\vartheta), y(\vartheta)) = 2\int_{0}^{x} f(\xi) \, d\xi + y^2 + \nu^2 \int_{-h}^{0} \left(\int_{\vartheta_1}^{0} y^2(\vartheta) \, d\vartheta\right) d\vartheta_1$$

and evaluate $\lim_{\Delta t \to +0}(\Delta V/\Delta t)$ along a trajectory of equation (34.16). The result is

(34.22)
$$\lim_{\Delta t \to +0}\left(\frac{\Delta V}{\Delta t}\right)_{(34.16)} = -2y\varphi(y, t) + 2\int_{-h(t)}^{0} f'(x(t + \vartheta))y(t + \vartheta)y(t) \, d\vartheta$$
$$+ \nu^2 \int_{-h}^{0} (y^2(t) - y^2(t + \vartheta)) \, d\vartheta .$$

If we set $\nu = a/h$, the estimate

(34.23)
$$\lim_{\Delta t \to +0}\left(\frac{\Delta V}{\Delta t}\right)_{(34.16)} < -\int_{-h}^{0}\left(\frac{a}{h}y^2 - 2L \,|\, y(t)y(t + \vartheta) \,|\, + \frac{a}{h}y^2(t + \vartheta)\right) d\vartheta$$

is obtained, if we take account of the relations (34.17) and (34.18).

A sufficient condition that the functional V satisfy the hypotheses of theorem 31.3 is

(34.24)
$$h < \frac{a}{L} .$$

This is so since under condition (34.24) the function of the arguments $y(t)$ and $y(t + \vartheta)$ under the integral in the right member of inequality (34.23)

is positive-definite. Thus if conditions (34.23) are satisfied, the right derivative $\lim_{\Delta t \to +0}(\Delta V/\Delta t)$ evaluated along trajectories of the system (34.20) is nonpositive, and the relation $\lim_{\Delta t \to +0}(\Delta V/\Delta t) = 0$ can hold for all $t \geq h$ only if $y(t) = 0$. But the second equation of (34.20) shows that the identity $y(t) \equiv 0$ for $t \geq h$ can hold only for $x(t) \equiv 0$ for values $t > t_0 > 0$. It is thus seen that if relation (34.24) holds, the functional V does satisfy the hypotheses of theorem 31.3, and the null solution $x = 0$, $y = 0$ of equation 34.16 is asymptotically stable.

We next consider the first-order equation

$$(34.25) \qquad \frac{dx}{dt} = f(x(t - h(t)), t) \qquad (0 \leq h(t) \leq h) ,$$

where $f(x, t)$ is a continuous function of its arguments. We assume that the function $f(x, t)$ has a continuous partial derivative $\partial f/\partial x$ that satisfies the inequality

$$(34.26) \qquad \left| \frac{\partial f}{\partial x} \right| < L \qquad (L = \text{const.}) .$$

We rewrite equation (34.25) as follows: For $t > 2h$, we write

$$\frac{dx}{dt} = f(x, t) + [f(x(t - h(t)), t) - f(x, t)] ,$$

which can be further rewritten

$$(34.27) \qquad \frac{dx}{dt} = f(x, t) - \int_{t-h(t)}^{t} f'_x(x(\xi), t) f(x(\xi - h(\xi)), \xi) \, d\xi .$$

Let $v(x)$ be the function $v(x) = x^2$. The derivative dv/dt of this function along a trajectory of (34.27) is evaluated as follows:

$$\left(\frac{dv}{dt} \right)_{(34.27)} = 2x(t)f(x(t), t) - 2\int_{t-h(t)}^{t} x(t) f'_x(x(\vartheta), t) f(x(\vartheta - h(\vartheta)), \vartheta) \, d\vartheta .$$

From inequality (34.26) we get the further estimate

$$(34.28) \qquad \frac{dv}{dt} \leq 2x(t)f(x(t), t) + L^2 h(t) \,|\, x(t)x(t - \xi) \,| \quad \text{for } 0 < \xi < 2h .$$

A sufficient condition for the hypotheses of theorem 31.4 to be satisfied is that the right member of equality (34.28) be a negative-definite function on the curves

$$qx^2(t - \xi) \leq x^2(t) ,$$

where q is some fixed number between 0 and 1: $0 < q < 1$. A sufficient condition for this is that the functions $f(x, t)$, $h(t)$ satisfy the inequalities

$$(34.29) \qquad \frac{f(x, t)}{x} + L^2 h(t) < -\gamma \quad \text{for } \gamma > 0, \, x \neq 0 ,$$

where γ is a sufficiently small positive number. That is to say, inequality

(34.29) gives a condition sufficient for the null solution $x = 0$ of equation (34.25) to be asymptotically stable.

We now consider the second-order nonlinear equation

$$(34.30) \qquad \frac{d^2x}{dt^2} = X(x(t), y(t)) + \varphi(y(t - h), t)$$

$$\left(y = \frac{dx}{dt}; h \text{ is a positive constant}\right),$$

where the functions X and φ satisfy the requirements

$$(34.31) \qquad \frac{X(x, y) - X(x, 0)}{y} < -a , \qquad \frac{X(x, 0)}{x} < -b \qquad \text{for } x \neq 0, y \neq 0 ,$$

where a and b are positive constants,

$$(34.32) \qquad |\varphi(y, t)| \leq L |y| .$$

We write equation (34.30) in the equivalent form

$$(34.33) \qquad \begin{aligned} \frac{dx}{dt} &= y , \\ \frac{dy}{dt} &= X(x, 0) + [X(x, y) - X(x, 0)] + \varphi(y(t - h), t) . \end{aligned}$$

We define the functional V by

$$(34.34) \qquad V(x(\vartheta), y(\vartheta)) = -\int_0^x X(\xi, 0)\, d\xi + \frac{y^2}{2} + \frac{a}{2} \int_{-h}^0 y^2(\xi)\, d\xi .$$

To estimate the derivative dV/dt along a trajectory of the system (34.33), we write

$$\frac{dV}{dt} = [X(x, y) - X(x, 0)] y + y\varphi(y(t - h), t) + \frac{ay^2}{2} - \frac{ay^2(t - h)}{2} .$$

Conditions (34.31) and (34.32) give the estimates

$$(34.35) \qquad \frac{dV}{dt} \leq -\frac{ay^2(t)}{2} + L |y(t)y(t - h(t))| - \frac{ay^2(t - h)}{2} .$$

The functional V satisfies the hypotheses of theorem 31.3 if the right member of inequality (34.35) is a negative-definite function of the arguments $y(t)$ and $y(t - h)$ (section 31, p. 154). This condition is satisfied if the numbers a, L are related by

$$(34.36) \qquad a > L .$$

Thus inequality (34.36) is a condition sufficient for the asymptotic stability of the null solution $x = y = 0$ of equation (34.30), when written in the form (34.33).

References

The following abbreviations are used in this bibliography:

AT	Avtomatika i Telemehanika
DAN	Doklady Akademiï Nauk SSSR
PMM	Prikladnaya Matematika i Mehanika
UMN	Uspehi Matematičeskih Nauk
VMU	Vestnik Moskovskogo Universiteta
Gostehizdat	Gosudarstvennoe Izdatelstvo Tehno-teoretičeskoï Literatury, Moskva

[1] AĬZERMAN, M. A., Convergence of a Regulatory Process for Large Initial Disturbances, *AT*, **7** (1946), 2-3.

[2] ———, Use of Nonlinear Functions of Several Arguments to Determine Stability of a Servomechanism, *AT*, **8**, 1 (1947).

[3] ———, A Problem Connected with the Stability of Large Dynamical Systems, *UMN*, **4**, 4 (1947).

[4] ———, A Sufficient Condition for the Stability of a Class of Dynamical Systems with Variable Parameters, *PMM*, **15**, 3 (1951).

[5] ———, *Theory of Servomechanisms*. Gostehizdat, 1952.

[6] ———, and F. R. GANTMAHER, The Determination of Stability by Means of a Linear Approximation of Periodic Solutions of Simultaneous Differential Equations with Discontinuous Right-Hand Sides, *DAN*, **116**, 4 (1957).

[7] ALEKSANDROV, P. S., *Introduction to the General Theory of Sets and Functions*. Gostehizdat, 1948.

[8] AMINOV, M. S., A Method for Obtaining Sufficient Conditions for Stability of Motion, *PMM*, **19**, 5 (1955).

[9] ANTOSIEWICZ, H. A., Stable Systems of Differential Equations with Integrable Perturbation Term, *J. London Math. Soc.*, **31** (1956), 208-12.

[10] ———, A Survey of Lyapunov's Second Method, in S. Lefschetz (ed.), *Contributions to the Theory of Nonlinear Oscillations*, Vol. IV (Annals of Mathematical Studies No. 41). Princeton, N.J.: Princeton Univ. Press, 1959, pp. 141-66.

[11] ARTEM'EV, N. A., Realizable Motion, *Izvestiya Akad. Nauk SSSR*, Ser. Mat., **3** (1939), 429-46.

[12] ATKINSON, F. V., On Stability and Asymptotic Equilibrium, *Ann. Math.* (2), **68** (1958), 690-708.

[13] BARBAŠIN, E. A., Existence of Smooth Solutions of Some Linear Equations with Partial Derivatives, *DAN*, **72**, 3 (1950).

[14] ————, Method of Sections in the Theory of Dynamical Systems, *Mat. Sbor.*, **29**, 2 (1951).

[15] ————, Stability of Solutions of a Nonlinear Equation of Third Order, *PMM*, **16**, 5 (1952).

[16] ————, and N. N. KRASOVSKIĬ, Stability of Motion in the Large, *DAN*, **86**, 3 (1952).

[17] ————, and N. N. KRASOVSKIĬ, Existence of Lyapunov Functions for Asymptotic Stability in the Large, *PMM*, **18**, 3 (1954).

[18] ————, and M. A. SKALKINA, A Question of First-Order Stability, *PMM*, **19**, 5 (1955).

[19] BELLMAN, R., *A Survey of the Theory of the Boundedness, Stability, and Asymptotic Behavior of Solutions of Linear and Nonlinear Differential and Difference Equations.* Washington, D.C.: Office of Naval Research, 1949.

[20] ————, On the Existence and Boundedness of Solutions of Nonlinear Differential-Difference Equations, *Ann. Math.*, **50**, 2 (1949).

[20a] ————, *Adaptive Control Processes; a Guided Tour.* Princeton, N. J.: Princeton Univ. Press, 1961.

[21] ————, and KENNETH L. COOKE, Asymptotic Behavior of Solutions of Differential-Difference Equations (*Mem. Amer. Math. Soc.* No. 35). Providence, R.I.: The Society, 1959.

[22] ————, and KENNETH L. COOKE, On the Limit of Solutions of Differential-Difference Equations as the Retardation Approaches Zero, *Proc. Nat. Acad. Sci. U.S.A.*, **45** (1959), 1026-28.

[23] BOGUSZ, WLADYSLAW, Determination of Stability Regions of Dynamic Non-linear Systems, *Arch. Mech. Stos.*, **11** (1959), 691-713 (Polish and Russian summaries).

[24] BOLTYANSKIĬ, V. G., and L. S. PONTRYAGIN, Stability of Positive Isobaric Rayleigh Systems of Ordinary Differential Equations, *Trudy 3rd mat. Cong.*, **1** (1956), 217.

[25] BRAUER, FRED, Global Behavior of Solutions of Ordinary Differential Equations, *J. Math. Anal. and Appl.*, **2** (1961), 145-58.

[25a] ————, Bounds for Solutions of Ordinary Differential Equations. Madison, Wis.: Mathematics Research Center Report No. 272, November 1961.

[25b] ————, Asymptotic Equivalence and Asymptotic Behavior of Linear Systems, *Michigan Math. J.*, **9** (1962), 33-43.

[26] ————, and S. STERNBERG, Local Uniqueness, Existence in the Large, and the Convergence of Successive Approximations, *Amer. J. Math.*, **80** (1958), 421-30; **81** (1959), 797.

[26a] BROWNELL, F. H., and W. K. ERGEN, A Theorem on Rearrangements and Its Application to Certain Delay Differential Equations, *J. Rat. Mech. Anal.*, **3** (1954), 565-79.

[27] CARTWRIGHT, M., Forced Oscillations in Nonlinear Systems, in S. Lefschetz (ed.), *Contributions to the Theory of Nonlinear Oscillations*, Vol. I (Annals of Mathematics Studies, No. 20). Princeton, N.J.: Princeton Univ. Press, 1950.

[27a] ————, On the Stability of Solutions of Certain Differential Equations of the Fourth Order, *Quart. J. Mech.*, **9** (1956), 185-94.

[28] CESARI, LAMBERTO, *Asymptotic Behavior and Stability Problems in Ordinary Differential Equations.* Berlin: Springer, 1959.

[29] ČETAEV, N. G., The Smallest Proper Value, *PMM*, **9** (1945), 193-96.

[30] ————, A Problem of Cauchy, *PMM*, **9** (1945), 139-42.

[31] ————, A Question of Parametric Stability of a Mechanical System, *PMM*, **15** (1951), 371-72.

[32] ———, Instability of Equilibrium in Some Cases When the Forcing Function Is Not Maximal, *PMM*, **16** (1952), 89–93.

[33] ———, Stability of Rotation of a Rigid Body with a Fixed Point in the Lagrange Case, *PMM*, **18** (1954), 123–24.

[34] ———, A Property of the Poincaré Equation, *PMM*, **19** (1955), 513–15.

[35] ———, Some Problems of Stability of Motion in Mechanics, *PMM*, **20** (1956), 309–14.

[36] ———, *Stability of Motion* (2nd ed.). Gostehizdat, 1956.

[37] CHIN YUAN-SHUN, LIU YING-CHING, and WANG LIAN, On the Equivalence Problem of Differential Equations and Difference-Differential Equations in the Theory of Stability, *Acta Math. Sinica*, **9** (1959), 333–63 (Chinese; English summary).

[38] ———, Unconditional Stability of Systems with Time-Lags, *Acta Math. Sinica*, **10** (1960), 125–42 (Chinese; English summary).

[39] CODDINGTON, E. A., and N. LEVINSON, *Theory of Ordinary Differential Equations*. New York: McGraw-Hill, 1955.

[39a] COURANT, R., *Differential and Integral Calculus*. New York: Interscience. Vol. I, 1937; Vol. II, 1936.

[39b] DRIVER, R., Existence and Stability of Solutions of a Delay-Differential System, *Archive for Rational Mechanics and Analysis*, **10** (1962), 401–26.

[40] DUBOŠIN, G. N., The Question of Stability of Motion for Persistent Disturbances, *Trudy Astr. Inst. im. Sternberg*, **14**, 1 (1940).

[41] DUVAKIN, A. P., and A. M. LETOV, Stability of Regulatory Systems with Two Degrees of Freedom, *PMM*, **18**, 2 (1954).

[42] ÈL'SGOL'C, L. È., *Qualitative Methods in Mathematical Analysis*. Gostehizdat, 1955.

[43] ———, Stability of the Solution of a Differential-Difference Equation, *UMN*, **9**, 4 (1954).

[44] ———, On the Theory of Stability of Differential Equations with Retarded Argument, *VMU*, No. 5 (1959), 65–71.

[45] ———, Some Properties of Periodic Solutions of Linear and Quasi-Linear Differential Equations with Retarded Argument, *VMU*, No. 5 (1959), 229–34.

[46] ERGEN, W. K., H. J. LIPKIN, and J. A. NOHEL, Applications of Liapunov's Second Method in Reactor Dynamics, *J. Math. Phys.*, **36** (1957), 36–48.

[47] ERŠOV, B. A., Stability in the Large of Some Systems of Servomechanisms, *PMM*, **17**, 1 (1953).

[48] ———, A Theorem on Stability of Motion in the Large, *PMM*, **18**, 3 (1954).

[49] ERUGIN, N. P., Some Questions in the Theory of Stability of Motion and Qualitative Theory of Differential Equations, *PMM*, **14**, 5 (1950).

[50] ———, Qualitative Determination of the Integral Curves of a System of Differential Equations, *PMM*, **14**, 6 (1950).

[51] ———, A Problem in the Theory of Stability of Servomechanisms, *PMM*, **16**, 5 (1952).

[52] ———, Qualitative Methods in the Theory of Stability, *PMM*, **19**, 5 (1955).

[53] ———, Methods for Solution of the Stability Problem in the Large, *Transactions of the Second All-Union Congress on the Theory of Servomechanisms*, Vol. I. Moscow-Leningrad Izdat. Akad. Nauk SSSR, 1955.

[54] ———, Periodic Solutions of Differential Equations, *PMM*, **20**, 1 (1956).

[55] FIHTENGOL'C, G. M., *Course in Differential and Integral Calculus*, Vol. I. Gostehizdat, 1948.

[56] ———, *Course in Differential and Integral Calculus*, Vol. II. Gostehizdat, 1949.

[57] FILIMONOV, YU. M., On the Stability of Solutions of Differential Equations of Second Order, *PMM*, **23** (1959), 596-98 (Russian) [translated as *J. Appl. Math. Mech.*, **23** (1959), 846-49].

[58] FILIPPOV, A. F., On Continuous Dependence of a Solution on the Initial Conditions, *UMN*, **14** (1959), 6 (90), 197-201.

[59] GERMAÏDZE, V. E., First-Order Asymptotic Stability, *PMM*, **21**, 1 (1957).

[60] ———, and N. N. KRASOVSKIĬ, Stability for Persistent Disturbances, *PMM*, **21**, 6 (1957).

[60a] GANTMACHER, F. R., *Applications of the Theory of Matrices*. New York: Interscience, 1959.

[61] GORBUNOV, A. D., Conditions for Asymptotic Stability of the Null Solution of Systems of Ordinary Linear Homogeneous Differential Equations, *VMU*, No. 9 (1953), 49-55.

[62] GORELIK, G. S., Theory of Inverse Connection, *J. Teh. Phys.*, **50** (1939).

[63] GORŠIN, S. I., Stability of Motion for Persistent Disturbances, *Izvestiya Akad. Nauk Kazah. SSR*, **56**. Set. Mat. Meh. **2**, 46-73 (1948).

[64] ———, Stability of Solutions of a Pair of Systems of Differential Equations with Persistent Disturbances, *ibid.*, **60** (**3**), 32-38 (1949).

[65] ———, Lyapunov's Second Method, *ibid.*, **97** (**4**), 42-50 (1950).

[65a] ———, Some Criteria of Stability under Persistent Disturbances, *ibid.*, **97** (**4**), 51-56 (1950).

[65b] GRAVES, L. M., Some General Approximation Theorems, *Ann. Math.* (**2**), **42** (1941), 281-92.

[66] HAHN, W., Über Stabilität bei nichtlinearen Systemen, *Z. angew. Math. Mech.*, **35**, 12 (1955).

[67] ———. Bemerkungen zu einer Arbeit von Herrn Vejvoda über Stabilitätsfragen, *Math. Nachr.*, **20** (1959), 21-24.

[68] ———, *Theorie und Anwendung der direkten Methode von Ljapunov*. Berlin: Springer, 1959.

[69] HALANAĬ, A. (HALANAY, A.), An Averaging Method for Systems of Differential Equations with Lagging Argument, *Rev. Math. Pures Appl.*, **4** (1959), 467-83.

[70] ———, Solutions périodiques des systèmes linéaires à argument retardé, *C. R. Acad. Sci. Paris*, **249** (1959), 2708-2709.

[71] ———, Sur les systèmes d'équations différentielles linéaires à argument retardé, *C. R. Acad. Sci. Paris*, **250** (1960), 797-98.

[71a] HALE, J. K., Asymptotic Behavior of the Solutions of Differential-Difference Equations. Tech. Rep. 61-10, RIAS, Baltimore, 1961. Proc. International Symposium on Nonlinear Vibrations, Kiev (1961), 1963.

[72] HARASAHAL, V., First-Order Stability of Solutions of Countable Systems of Differential Equations, *Izvestiya Akad. Nauk Kazah. SSR*, **60**, 3 (1949).

[72a] HARTMAN, P., On Stability in the Large for Systems of Ordinary Differential Equations, *Canadian J. Math.*, **13** (1961), 480-92.

]72b] ———, and C. OLECH, *On Global Asymptotic Stability of Solutions of Differential equations*. The Johns Hopkins University, 1962.

[73] HILLE, E., and R. S. PHILLIPS, *Functional Analysis and Semi-Groups* (Colloquium Publications, Vol. 31). New York: American Mathematical Society, 1956.

[74] HUKUHARA, MASUO, and MASAHIRO IWANO, Étude de la convergence des solutions formelles d'un système différentiel ordinaire linéaire, *Funkcialaj Ekvacioj*, **2** (1959), 1-18 (Esperanto summary).

[75] KAKUTANI, S., and L. MARKUS, On the Non-linear Difference-Differential Equation $y'(t) = [A - By(t)]y(t)$, in S. Lefschetz (ed.), *Contributions to the Theory of Nonlinear Oscillations*, Vol. IV (Annals of Mathematical Studies No. 41). Princeton, N.J.: Princeton Univ. Press, 1959.

[76] KAMENKOV, G. V., Stability of Motion, *Sbor. Trudy Kazan Aviats. Inst.*, No. 9 (1939).

[77] ———, Stability of Motion in a Finite Interval of Time, *PMM*, **17**, 5 (1953).

[78] ———, and A. A. LEBEDEV, Remarks on the Paper on Stability in a Finite Interval of Time, *PMM*, **18**, 4 (1954).

[79] KARASEVA, T. M., On an "Exact Estimate" of the Multipliers of Second-Order Differential Equations with Periodic Coefficients, *DAN*, **121** (1958), 34–36.

[80] KIM, ČE DŽEN, Determination of the Region of Influence of an Asymptotically Stable Position of Equilibrium, *VMU*, No. 2 (1959), 3–14.

[81] KRASOVSKIĬ, N. N., Theorems on Stability of Motion Determined by a System of Two Equations, *PMM*, **16**, 5 (1952).

[82] ———, Stability of the Solution of a System of Second Order in the Critical Case, *DAN*, **93**, 6 (1953).

[83] ———, Stability of the Solution of a System of Two Differential Equations, *PMM*, **17**, 6 (1953).

[84] ———, The General Theorems of A. M. Lyapunov and N. G. Četaev for Stationary Systems of Differential Equations, *PMM*, **18**, 5 (1954).

[85] ———, Sufficient Conditions for Stability of Motion of Systems of Nonlinear Differential Equations, *DAN*, **98**, 6 (1954).

[86] ———, Stability of Motion in the Critical Case with a Single Zero Root, *Mat. Sbor.*, **37**, 1 (1955).

[87] ———, Transformation of the Theorem of K. P. Persidskiĭ on Uniform Stability, *PMM*, **19**, 3 (1955).

[88] ———, First-Order Stability, *PMM*, **19**, 5 (1955).

[89] ———, Theory of A. M. Lyapunov's Second Method for Investigating Stability, *Mat. Sbor.*, **40**, 1 (1956).

[90] ———, Examples of Lyapunov's Method for Stability with Time Delay, *PMM*, **20**, 2 (1956).

[91] ———, Transformation of the Theorem of A. M. Lyapunov's Second Method and Questions of First-Order Stability of Motion, *PMM*, **20**, 2 (1956).

[92] ———, Asymptotic Stability for Systems and Complements, *PMM*, **20**, 3 (1956).

[93] ———, Theory of Lyapunov's Second Method for Investigating Stability of Motion, *DAN*, **109**, 3 (1956).

[94] ———, Stability for Large Initial Perturbations, *PMM*, **21**, 3 (1957).

[95] KREĬN, M. G., Some Questions Connected with Lyapunov's Circle of Ideas on the Theory of Stability, *UMN*, **1**, 3 (1948).

[96] KURCVEĬL', J., Transformation of Lyapunov's First Theorem on Stability of Motion, *Czech. Mat. J.*, **5** (**80**) (1955), 382–98.

[97] ———, Transformation of Lyapunov's Second Theorem on the Stability of Motion, *Czech. Mat. J.*, **6** (**81**), 2 (1956); 4 (1956).

[98] ———, Generalized Ordinary Differential Equations and Continuous Dependence on a Parameter, *Czech. Mat. J.*, **7** (**82**) (1957), 418–99.

[99] ———, and ZDENĚK VOREL, Continuous Dependence of Solutions of Differential Equations on a Parameter, *Czech. Mat. J.*, **7** (**82**) (1957), 568–83.

[100] ———, and I. VRKOČ, Transformation of Lyapunov's Theorems on Stability and Persidskiĭ's Theorems on Uniform Stability, *Czech. Mat. J.*, **7** (**82**), 2 (1957).

[101] KUZMAK, G. E., Asymptotic Solutions of Nonlinear Second-Order Differential Equations with Variable Coefficients, *PMM*, **23** (1959), 515–26 (Russian) [translated as *J. Appl. Math. Mech.*, **23** (1959), 730–74].

[102] KUZ'MIN, P. A., Stability for Parametric Perturbations, *PMM*, **21**, 1 (1957).

[103] LASALLE, J. P., The Extent of Asymptotic Stability, *Proc. Nat. Acad. Sci. U.S.A.*, **46** (1960), 363–65.

[104] ———, and SOLOMON LEFSCHETZ, *Stability of Liapunov's Direct Method with Applications*. New York: Academic Press, 1961.

[105] LEBEDEV, A. A., The Problem of Stability of Motion for a Finite Interval of Time, *PMM*, **18**, 1 (1954).

[106] ———, A Method for Constructing Lyapunov Functions, *PMM*, **21**, 1 (1957).

[107] LEE SHEN-LING, Topological Structure of Integral Curves of the Differential Equation

$$y' = \frac{a_0 x^n + a_1 x^{n-1} y + \cdots + a_n y^n}{b_0 x^n + b_1 x^{n-1} y + \cdots + b_n y^n} ,$$

Acta Math. Sinica, **10** (1960), 1–21 (Chinese; English summary).

[108] LETOV, A. M., Stability of Systems of Automatic Regulators with Two Active Devices, *PMM*, **17**, 4 (1953).

[109] ———, *Stability of Nonlinear Systems of Automatic Regulators*. Gostehizdat, 1955.

[110] ———, Stability of Unstable Motion of Systems of Regulators, *PMM*, **19**, 3 (1955).

[110a] LEVIN, J. J., On the Global Asymptotic Behavior of Non-linear Systems of Differential Equations, *Archive for Rational Mechanics and Analysis*, **6** (1960), 65–74.

[111] LEVIN, J. J., and J. A. NOHEL, On a System of Integro-differential Equations Occurring in Reactor Dynamics, *J. Math. Mech.*, **9**, 3 (1960), 347–68; Abstract 581–13, *Not. Amer. Math. Soc.*, **8**, 3, Issue No. 54 (1961), 250.

[111a] ———, On a System of Integrodifferential Equations Occurring in Reactor Dynamics, II, *Archive for Rational Mechanics and Analysis*, **11** (1963).

[111b] ———, Global Asymptotic Stability for Nonlinear Systems of Differential Equations and Applications to Reactor Dynamics, *Archive for Rational Mechanics and Analysis*, **5** (1960), 194–211. (See also *Boletin de la Sociedad Matematica Mexicana* (1960), 152–57.)

[111c] ———, Note on a Nonlinear Volterra Equation, *Proceedings of the American Math. Soc.* To be published 1963.

[111d] ———, The Asymptotic Behavior of the Solution of a Volterra Equation, *Proceedings of the American Math. Soc.*, **13** (1963).

[111e] LEVIN, J. J., and J. A. NOHEL, On a Nonlinear Delay Equation. To appear.

[112] LUR'E, A. I., and V. N. POSTNIKOV, Theory of Stability of Servomechanisms, *PMM*, **8**, 3 (1944).

[113] ———, and V. N. POSTNIKOV, *Some Nonlinear Problems in the Theory of Automatic Regulation*. Gostehizdat, 1951.

[114] LYAPUNOV [LIAPOUNOFF], A. M., *Problème générale de la stabilité du mouvement* (Annals of Mathematical Studies, No. 17). Princeton, N.J.: Princeton Univ. Press, 1947.

[115] LYAŠČENKO, N. YA., The Problem of Asymptotic Stability of the Solution of Nonlinear Systems of Differential Equations, *DAN*, **104**, 2 (1955).

[116] MALKIN, I. G., The Problem of Existence of Lyapunov Functions, *Izvestiya Kazan. Filial Akad. Nauk SSSR, Ser. Fiz.-Mat. Teh. Nauk*, **5** (1931).

[117] ———, First-Order Stability, *Collected Scientific Works of Kazan Aviation Institute*, No. 3 (1935).

[118] ———, Some Fundamental Theorems in the Theory of Stability of Motion in Critical Cases, *PMM*, **6** (1942), 411–48.

[119] ———, Stability for Persistent Disturbances, *PMM*, **8** (1944), 327–31.

[120] ———, Theory of Stability of Servomechanisms, *PMM*, **15** (1951), 575–90.

[121] ———, Theorems on First-Order Stability, *DAN*, **86**, 6 (1951).

[122] ———, A Problem in the Theory of Stability of Systems of Automatic Regulators, *PMM*, **16** (1952), 365–68.

[123] ———, Stability of Systems of Automatic Regulators, *PMM*, **16** (1952), 495–99.

[124] ———, *Theory of Stability of Motion*. Gostehizdat, 1952.

[125] ———, Questions Concerning Transformation of Lyapunov's Theorem on Asymptotic Stability, *PMM*, **18** (1954), 129–38.

[126] ———, [Joel G.] *Theorie der Stabilität einer Bewegung*. Translation, ed. W. Hahn and R. Reissig. Munich: R. Oldenbourg, 1959.

[126a] MANGASARIAN, O. L., *Stability Criteria for Nonlinear Ordinary Differential Equations*. Shell Development Company, Emeryville, Calif., Report P-1146, 1962.

[127] MARCUS, M., Some Results on the Asymptotic Behaviour of Linear Systems, *Can. J. Math.*, **7** (1955), 531–38.

[128] MARKUS, L., and H. YAMABE, Global Stability Criteria for Differential Systems, *Osaka Math. J.*, **12** (1960), 305–17.

[129] MASSERA, J. L., On Liapounoff's Conditions of Stability, *Ann. Math.* (2), **50**, 3 (1949), 705–21.

[130] ———, Estabilidad total y vibraciones aproximadamente periodicas, *Publ. Inst. Mat. Estadistica*, **2**, 7 (1954).

[131] ———, Contributions to Stability Theory, *Ann. Math.*, **64**, 1 (1956), 182–206.

[132] ———, and J. J. SCHAFFER, Linear Differential Equations and Functional Analysis, III. Lyapunov's Second Method in the Case of Conditional Stability, *Ann. Math.* (2), **69** (1959), 535–74.

[133] MIKOLAJSKA, Z., Sur la stabilité asymptotique des solutions d'un systéme d'équations différentielles, *Ann. Polon. Math.*, **7** (1959), 13–19.

[134] MINORSKY, N., Control Problems, *J. Franklin Inst.*, **232**, 6 (1941).

[135] MOVŠOVIČ, S. M., Determination of the Domain of Attraction of Some Singular Points of Higher Order, *VMU*, No. 6 (1959), 3–11.

[135a] MUFTI, I. H., Stability in the Large of Systems of Two Equations, *Arch. Rational Mech. Analysis*, **7** (1961), 119–34.

[136] MYŠKIS, A. D., *Linear Differential Equations with Delay in the Argument*. Gostehizdat, 1951.

[136a] NATANSON, I. P., *Konstruktive Funktionentheorie*. Akademie Verlag, Berlin, 1955. Chap. I.

[137] NEMYCKIĬ, V. V., Fixed-point Methods in Analysis, *UMN*, **1** (1936), 141–74.

[138] ———, and V. V. STEPANOV, *Qualitative Theory of Differential Equations*. Translated under direction of S. Lefschetz. Princeton Univ. Press, 1960.

[139] ———, Estimates of the Region of Asymptotic Stability for Nonlinear Systems, *DAN*, **101** (1955), 803–4.

[140] ———, On Certain Methods of Qualitative Investigation "in the Large" of Many-Dimensional Autonomous Systems, *Trudy Moskov. Mat. Obs.*, **5** (1956), 455–82.

[141] NOHEL, JOHN A., A Class of Nonlinear Delay Differential Equations, *J. Math. Phys.*, **38** (1960), 295–311.

[141a] ———, Stability of Perturbed Periodic Motions, *J. reine angew. Math.*, **203** (1960), 64–79.

[141b] ———, Some Problems in Nonlinear Volterra Integral Equations, *Bull. Amer. Math. Soc.*, **68** (1962), 323–29.

[141c] OGURCOV, A. I., On the Stability of the Solutions of Certain Nonlinear Differential Equations of Third and Fourth Order, *Izv. Vysš. Učebn. Zaved Matematika*, 1959, no. 3 (10), pp. 200–209.

[142] OLECH, C., Estimates of the Exponential Growth of Solutions of a Second-Order Ordinary Differential Equation, *Bull. Acad. Polon. Sci., Ser. Sci. Math. Astr. Phys.*, **7** (1959), 487–94 (Russian summary; unbound insert).

[143] OPIAL, Z., Sur la stabilité des solutions périodiques et presque-périodiques de l'équation différentielle $x'' + F(x') + g(x) = p(t)$, *Bull. Acad. Polon. Sci., Ser. Sci. Math. Astr. Phys.*, **7** (1959), 495–500 (Russian summary; unbound insert).

[144] OSTROWSKI, A., and H. SCHNEIDER, Some Theorems on the Inertia of General Matrices. Madison, Wis.: Mathematics Research Center, 1961.

[145] PERSIDSKIĬ, K. P., Theory of Stability of Integrals of a System of Differential Equations, *Izv. fiz.-mat. pri Kazan. Gos. univ.*, Vol. VIII (1936–37).

[146] ———, A Theorem of Lyapunov, *DAN*, **14**, (1937).

[147] ———, Theory of Stability of Solutions of Differential Equations. Dissertation, 1946.

[148] ———, Theory of Solutions of Differential Equations, *Izvestiya Akad. Nauk Kazah. SSR*, **97**, 4 (1950).

[149] PETROVSKIĬ, I. G., *Lectures on the Theory of Ordinary Differential Equations.* Gostehizdat, 1947.

[150] PLISS, V. A., Qualitative Graphing of Integral Curves in the Large, and Construction of Arbitrary Points in the Region of Stability of a System of Two Differential Equations, *PMM*, **17** (1953), 541–54.

[151] ———, Necessary and Sufficient Conditions for Stability for Systems of n Differential Equations, *DAN*, **103** (1955), 17–18.

[152] ———, Investigation of a Nonlinear System of Three Differential Equations, *DAN*, **117** (1957), 184–87.

[153] ———, The Necessary and Sufficient Conditions for the Stability in the Large of a Homogeneous System of Three Differential Equations, *DAN*, **120** (1958), 708–10.

[154] RAZUMIHIN, B. S., Second-Order Stability of the Trivial Solution, *PMM*, **19** (1955), 279–86.

[155] ———, Stability of Motion, *PMM*, **20** (1956), 266–70.

[156] ———, Stability of Systems with Time Delay, *PMM*, **20** (1956), 500–512.

[157] ———, Stability of a System of Automatic Regulators with a Single Active Device, *AT*, **17**, 11 (1956).

[158] ———, Estimates for the Solution of a System of Differential Equations with Variable Coefficients, *PMM*, **21** (1957).

[159] ———, On the Application of the Lyapunov Method to Stability Problems, *PMM*, **22**, 3 (1958), 338–49 [translated as *J. Appl. Math. Mech.*, **22** (1958), 466–80].

[160] REGHIS, M., Sur la stabilité "d'après la première approximation," *Lucrăr. Sti. Inst. Ped. Timisoara. Mat.-Fiz.* (1959), 135–44 (Romanian; French and Russian summaries).

[161] REISSIG, ROLF, Kriterien für die Zugenhörigkeit dynamischer Systeme zur Klasse, *Math. Nachr.*, **20** (1959), 67–72.

[162] ———, Ein Kriterium für asymptotische Stabilität, *Z. angew. Math. Mech.*, **40** (1960), 94–99.

[163] REPIN, YU. M., On Stability of Solutions of Equations with Retarded Argument, *PMM*, **21** (1957), 253–61.

[164] RIESZ, F., and B. SZ-NAGY, *Lectures on Functional Analysis* (2nd ed.). New York: Ungar, 1955.

[165] ROĬTENBERG, YA. N., On the Accumulation of Perturbations in Non-Stationary Linear Systems, *DAN*, **121** (1958), 221–24.

[166] RUMYANCEV, V. V., S. A. Calygin's Conditions for the Stability of Helical Motion of a Rigid Body in a Fluid, *PMM*, **19** (1955), 229–30.

[167] ———, Stability of Permanent Rotation of a Heavy Rigid Body, *PMM*, **20** (1956), 51–66.

[168] ———, Theory of Stability of Servomechanisms, *PMM*, **20** (1956), 714–22.

[169] ———, Stability of Permanent Motion of a Rigid Body Around a Fixed Point, *PMM*, **21** (1957), 339–46.

[170] ŠIMANOV, S. N., Stability of the Solution of a Nonlinear Equation of Third Order, *PMM*, **17** (1953), 369–72.

[171] ———, On the Instability of the Motion of Systems with Retardation, *PMM*, **24** (1960), 55–63 [translated as *J. Appl. Math. Mech.*, **24** (1960), 70–81].

[172] SKAČKOV, B. N., Stability in the Large of a Class of Nonlinear Systems of Automatic Regulators, *Vestnik Leningr. Univ.*, No. 1 and No. 13 (1957).

[173] SKALKINA, M. A., Preservation of Asymptotic Stability When a Differential Equation is Changed to the Corresponding Discrete Equation, *DAN*, **104** (1955), 505–8.

[173a] SOLODOVNIKOV, V. V., *Introduction to the Statistical Dynamics of Automatic Control Systems*. Translation, ed. J. B. Thomas and L. A. Zadeh. New York: Dover, 1960 (1952).

[174] SPASSKIĬ, R. A., A Class of Regulatory Systems, *PMM*, **18** (1954), 329–44.

[175] STARŽINSKIĬ, V. M., Sufficient Conditions for Stability of a Mechanical System with One Degree of Freedom, *PMM*, **16** (1952), 369–74.

[176] ———, Stability of a Mechanical System with One Degree of Freedom, *PMM*, **17** (1953), 117–22.

[177] ———, Stability of Motion in a Special Case, *PMM*, **19** (1955), 471–80.

[177a] TIMLAKE, W. P., and J. A. NOHEL, *Further Results on Global Behavior of Nonlinear Systems Occurring in Reactor Dynamics*. Oak Ridge National Laboratory, Report 60-10-50, October 17, 1960.

[178] TSIEN, H. S., *Engineering Cybernetics*. New York: McGraw-Hill, 1954.

[179] TUZOV, A. P., Questions of Stability for a System of Regulators, *Vestnik Leningr. Univ.*, No. 2 (1955).

[180] VEDROV, V. S., Stability of Motion, *Trudy Cent. Aero-Gidrodynam. Inst.* (1937), 327.

[181] VEKSLER, D. (WEXLER, D.), Stability theorems for a system of stationary differential equations, *Rev. Math. Pures Appl.*, **3** (1958), 131–38 (Russian).

[182] VINOGRAD, R. E., Remarks on the Critical Case of Stability of a Singular Point in the Plane, *DAN*, **101**, 2 (1955).

[183] VRKOČ, I., A General Theorem of Četaev, *Czech. Mat J.*, **5** (**80**) (1955).

[184] WANG, LIAN, On the Equivalence Problem of Differential Equations and Difference-Differential Equations in the Theory of Stability of the First Critical Case, *Acta Math. Sinica*, **10** (1960), 104–24 (Chinese; English summary).

[185] WAŻEWSKI, T., Sur la limitation des intégrales des systèmes d'équations différentielles linéaires ordinaires, *Studia Math.*, **10** (1948), 48–59.

[186] ———, Sur un problème asymptotique relatif au système de deux équations différentielles ordinaires, *Ann. Mat. Pura Appl.* (4), **49** (1960), 139–46.

[186a] WHITTAKER, E. T., and G. N. WATSON, *A Course of Modern Analysis* (4th ed.). Cambridge, Eng.: Cambridge Univ. Press, 1927; reprinted 1952.

[187] WRIGHT, E. M., The Linear Difference-Differential Equations with Asymptotically Constant Coefficients, *Amer. J. Math.*, **70** (1948), 221–38.

[188] ———, The Stability of Solutions of Nonlinear Difference-Differential Equations, *Proc. Roy. Soc.* (Edinburgh), **63**, 1 (1950).

[189] YAKUBOVIČ, V. A., A Class of Nonlinear Differential Equations, *DAN*, **117**, 1 (1957).

[190] ———, Stability in the Large of Unperturbed Motion for Equations of Indirect Automatic Regulators, *Vestnik Leningr. Univ.*, No. 19 (1957).

[191] ———, On Boundedness and Stability in the Large of the Solutions of Some Nonlinear Differential Equations, *DAN*, **121** (1958), 984–86.

[192] YOSHIZAWA, T., On the Stability of Solutions of a System of Differential Equations, *Mem. Coll. Sci. Univ. Kyoto, Ser. A. Math.*, **29** (1955), 27–33.

[193] ———, Liapunov's Function and Boundedness of Solutions, *Funkcialaj Ekvacioj*, **2** (1959), 95–142.

[194] ———, On the Equiasymptotic Stability in the Large, *Mem. Coll. Sci. Univ. Kyoto, Ser. A. Math.*, **32** (1959), 171–80.

[195] ZUBOV, V. I., Sufficient Conditions for Stability of Nonlinear Systems of Differential Equations, *PMM*, **17**, 4 (1953).

[196] ———, Theory of Lyapunov's Second Method, *DAN*, **99**, 3 (1954).

[197] ———, Theory of A. M. Lyapunov's Second Method, *DAN*, **100**, 5 (1955).

[198] ———, Questions in the Theory of Lyapunov's Method for Constructing a General Solution for Asymptotic Stability in the Plane, *PMM*, **19**, 6 (1955).

[199] ———, *Examples of A. M. Lyapunov's Methods*. Izdat. Leningr. Univ., 1957.

[200] ———, Some Problems in Stability of Motion, *Mat. Sbor.*, **48** (**90**) (1959), 149–90.

[201] ———, On Almost Periodic Solutions of Systems of Differential Equations, *Vestnik Leningr. Univ.*, No. 15 (1960), 104–6 (Russian; English summary).

[202] ZVERKIN, A. M., Dependence of the Stability of Solutions of Linear Differential Equations with Lagging Argument upon the Choice of the Initial Moment, *VMU*, No. 5 (1959), 15–20.

Index